ON HEIDEGGER

BEING AND TIME

On Heidegger's Being and Time is an outstanding exploration of Heidegger's most important work by two major philosophers. Simon Critchley argues that we must see *Being and Time* as a radicalization of Husserl's phenomenology, particularly his theories of intentionality, categorial intuition and the phenomenological concept of the a priori. This leads to a reappraisal and defense of Heidegger's conception of phenomenology.

In contrast, Reiner Schürmann urges us to read Heidegger "backward", arguing that his later work is the key to unravelling *Being and Time*. Through a close reading of *Being and Time* Schürmann demonstrates that this work is ultimately aporetic because the notion of Being elaborated in his later work is already at play within it. This is the first time that Schürmann's renowned lectures on Heidegger have been published.

The book concludes with Critchley's reinterpretation of the importance of authenticity in *Being and Time*. Arguing for what he calls an "originary inauthenticity", Critchley proposes a relational understanding of the key concepts of the second part of *Being and Time*: death, conscience and temporality.

Simon Critchley is Professor and Chair of Philosophy at the New School for Social Research in New York. His books include *The Ethics of Deconstruction, Things Merely Are: Philosophy in the Poetry of Wallace Stevens, Very Little...Almost Nothing* (Second edition) and *Infinitely Demanding: Ethics of Commitment, Politics of Resistance*.

Reiner Schürmann was, until his death in 1993, Professor of Philosophy at the New School for Social Research. His books include *Heidegger on Being and Acting* and *Broken Hegemonies*.

Steven Levine is Assistant Professor of Philosophy at the University of Massachusetts, Boston. He is the author of many articles on Contemporary Pragmatism and Critical Theory.

ON HEIDEGGER'S
BEING AND TIME

Simon Critchley and Reiner Schürmann

Edited by Steven Levine

Routledge
Taylor & Francis Group

LONDON AND NEW YORK

First published 2008 by Routledge
2 Park Square, Milton Park, Abingdon, Oxon OX14 4RN

Simultaneously published in the USA and Canada
by Routledge
270 Madison Ave, New York, NY 10016

Routledge is an imprint of the Taylor & Francis Group, an informa business

Typeset in Times by
GreenGate Publishing Services, Tonbridge, Kent

Printed and bound in Great Britain by
TJ International, Padstow, Cornwall

British Library Cataloguing in Publication Data
A catalogue record for this book is available from the British Library

Library of Congress Cataloging in Publication Data
Schürmann, Reiner, 1941–
On Heidegger's Being and time/Simon Critchley and Reiner Schürmann;
edited by Steven Levine.
p. cm.
Includes bibliographical references and index.
ISBN 978-0-415-77595-3 (hbk: alk. paper) – ISBN 978-0-415-77596-0
(pbk: alk. paper) 1. Heidegger, Martin, 1889–1976. Sein und Zeit. 2.
Ontology. 3. Space and time. I. Critchley, Simon, 1960– II. Levine,
Steven. III. Title.

B3279.H48S4676 2008
111–dc22
2007052705

ISBN 10: 0-415-77595-7 (hbk)
ISBN 10: 0-415-77596-5 (pbk)
ISBN 10: 0-415-46964-3 (ebk)

ISBN 13: 978-0-415-77595-3 (hbk)
ISBN 13: 978-0-415-77596-0 (pbk)
ISBN 13: 978-0-415-46964-7 (ebk)

CONTENTS

ILLUSTRATIONS

Figures

Tables

CONTRIBUTORS

Simon Critchley is Professor and Chair of Philosophy at the New School for Social Research in New York. He is author of many books, most recently *Infinitely Demanding* (2007) and *The Book of Dead Philosophers* (2008).

Steven Levine is an Assistant Professor of Philosophy at the University of Massachusetts, Boston. He is the author of many articles on Contemporary Pragmatism and Critical Theory.

Reiner Schürmann, was, until his death in 1993, Professor of Philosophy at the New School for Social Research. He is the author of many books, including *Heidegger on Being and Acting: From Principles to Anarchy* and *Broken Hegemonies*.

ABBREVIATIONS

The following abbreviations are used to refer to works by Heidegger in the text and notes:

BH "Brief über den 'Humanismus'", in *Wegmarken* (Frankfurt am Main: Vittorio Klostermann, 1967), pp. 145–94; "Letter on Humanism", in *Basic Writings*, trans. David F. Krell (San Francisco: HarperCollins, 1977), pp. 189–242; and *Pathmarks*, ed. William. McNeill (Cambridge: Cambridge University Press, 1998), pp. 239–76.

EM *Einführung in die Metaphysik* (Tübingen: Max Niemeyer, 1957); *An Introduction to Metaphysics*, trans. Ralph Manheim (New Haven: Yale University Press, 1968).

FD *Die Frage nach dem Ding* (Tübingen: Max Niemeyer, 1975). *What Is a Thing?*, trans. W. B. Barton Jr and Vera Deutsch (South Bend: Regnery/Gateway, 1967).

GA *Gesamtausgabe* (Frankfurt am Main: Vittorio Klostermann, 1975), vols 1–102.

GP *Die Grundprobleme der Phänomenologie* (GA 24, 1975). *The Basic Problems of Phenomenology*, trans. A. Hofstadter (Bloomington: Indiana University Press, 1982).

HCT *History of the Concept of Time: Prolegomena*, trans. T. Kisiel (Bloomington: Indiana University Press, 1985).

JS *Being and Time*, trans. Joan Stambaugh (Albany: SUNY, 1996).

KPM *Kant und das Problem der Metaphysik* (Frankfurt am Main: Vittorio Klostermann, 1973). *Kant and the Problem of Metaphysics*, trans. James S. Churchill (Bloomington: Indiana University Press, 1962); and *Kant and the Problem of Metaphysics*, trans. Richard Taft (Bloomington: Indiana University Press, 1997).

KTS *Kants These über das Sein* (Frankfurt am Main: Vittorio Klostermann, 1963); "Kant's Thesis About Being", (1973) trans. T. E. Klein and W. E. Pohl, *Southwestern Journal of Philosophy*, IV; 3, pp. 7–34; and *Pathmarks*, trans. W. McNeill (Cambridge: Cambridge University Press, 1998), pp. 337–63.

MR *Being and Time*, trans. J. Macquarrie and E. Robinson (New York: Harper & Row, 1979).

N *Nietzsche* (Pfullingen: Günther Neske, 1961), vols I and II; *The Will to Power as Art*, ed. and trans. David F. Krell (New York: Harper & Row, 1979); *The Eternal Recurrence of the Same*, ed. and trans. David F. Krell (New York: Harper & Row, 1984); *The Will to Power as Knowledge and Metaphysics* ed. David F. Krell, trans. Joan Stambaugh, David F. Krell and F. A. Capuzzi (New York: Harper & Row, 1984); *Nihilism*, ed. David F. Krell, trans. F. A. Capuzzi (New York: Harper & Row, 1982).

PGZ *Prolegomena zur Geschichte des Zeitbegriffs* (Frankfurt am Main: Vittorio Klostermann, 1979).

SD *Zur Sache des Denkens* (Tübingen: Max Niemeyer, 1969); *On Time and Being*, trans. Joan Stambaugh (New York: Harper & Row, 1972).

SG *Satz vom Grund* (Pfullingen: Günther Neske, 1957); *The Principle of Reason*, trans. R. Lilly (Bloomington: University of Indiana Press, 1977).

SZ Heidegger, *Sein und Zeit* (Tübingen: Max Niemeyer, 1984).

US *Unterwegs zur Sprache* (Pfullingen: Günther Neske, 1959); *On the Way to Language*, trans. P. D. Hertz (New York: Harper & Row, 1971).

WG "Vom Wesen des Grundes", in *Wegmarken* (Frankfurt am Main: Vittorio Klostermann, 1967); *The Essence of Reasons*, trans. Terrence Malick (Evanston: Northwestern University Press, 1969); and "On the Essence of Ground", in *Pathmarks*, ed. William McNeill (Cambridge: Cambridge University Press, 1998), pp. 337–64.

INTRODUCTION

Steven Levine

At the beginning of his essay "Heidegger for Beginners", Simon Critchley asks the question of where one should begin with Heidegger. The two authors of this book answer this question in different, yet complementary, ways. Critchley argues that we must read Heidegger "forwards" from phenomenology. To understand *Being and Time* and Heidegger's thinking overall we must see it as a radicalization of Husserl's phenomenological project. This thesis informs Critchley's essay "Heidegger for Beginners", in which he provides the essential phenomenological background necessary for understanding *Being and Time*. Reading *Being and Time* through Heidegger's 1925 lecture *History of the Concept of Time: Prolegomena*, Critchley shows how Heidegger's transformation of certain Husserlian themes—intentionality, categorial intuition, and the phenomenological concept of the a priori—makes sense of Heidegger's obscure master question: the question of the meaning of Being.

In contrast, Reiner Schürmann argues in his seminal book *Heidegger on Being and Acting: From Principles to Anarchy* that we must read Heidegger "backwards". By this he means that many of the key features and motivations for Heidegger's early work—especially as it concerns the question of the meaning of Being—can only be discerned by reading it in the light of the trajectory of his later work. This hermeneutical strategy puts great strain on the reader insofar as it requires one to have read all of Heidegger to understand *Being and Time*. Fortunately, in the lecture included in this book, Schürmann does much of the work for us by providing a detailed reading of *Being and Time* that is informed by his demanding hermeneutical strategy.

Both of these interpretive strategies break significantly with what is now the orthodox interpretation of Heidegger in the Anglo-American world. This orthodoxy, best represented by Hubert Dreyfus's seminal book *Being-In-the-World*, interprets Heidegger as a type of pragmatist.[1] On this interpretation, *Being and Time* is concerned above all with overcoming the philosophy of consciousness by replacing Husserl's transcendental phenomenology with a hermeneutical phenomenology. Husserl's theory focuses on intentionality, on the fact that conscious states are of something (an object or state of affairs) because they have intentional content. His theory therefore reproduces the Cartesian view that conscious experience takes the form of a subject immediately apprehending an inner representation of an outer object. Heidegger counters this view, Dreyfus claims, by demonstrating that there is "a more basic form of intentionality than that of a self-sufficient individual subject directed at the world by means of its mental content. At the foundation of Heidegger's new approach is a phenomenology of 'mindless' everyday coping skills as the basis of all intelligibility".[2]

Heidegger does not claim that the subject–object model that underlies the theory of knowledge is wrong, but rather that it is derivative upon a more primordial way of being-in-the-world. Agents can represent and know the world only because they already have in place an implicit understanding of the world, which is embedded in their coping practices. Subjects, for the most part, do not stand over against the world by representing it; rather they always already act within a meaningful world that they already in some sense understand.

While this picture of Heidegger is in some respects correct, for Schürmann and Critchley it is ultimately misleading. The first problem with it is that it trivializes the question of the meaning of Being. For Dreyfus this question simply asks about the understanding of being that is embedded in Dasein's pre-representational coping. Heidegger's fundamental ontology is thus concerned with "the nature of this understanding of being that we do not know—that is not a representation in the mind corresponding to the world—but that we simply *are*".[3]

The problem with this interpretation, according to Schürmann, is that it "does not operate with the understanding of the word 'Being' that Heidegger explicitly works out. 'Being' is *not* primarily man's, i.e. Dasein's being" (57).[4] Even in *Being and Time* Heidegger was "preoccupied with the question of Being as such—whatever that will turn out to mean—and only *therefore* with the question of Dasein" (57). Both the early and late Heidegger, Schürmann posits, are guided by the same question, the meaning of being as such. He thus contests the common notion that Heidegger, after *Being and Time*, makes a fundamental "turn" or a "reversal" from a consideration of Dasein's being to a consideration of the history of the epochs of Being.

Schürmann attempts to demonstrate the continuity of the early and later Heidegger through a close reading of the first eight sections of *Being and Time*. There, Schürmann claims, Heidegger lays out three intertwined projects that all contribute to a retrieval of the question of Being:

1   a transcendental analytic of Dasein, which demonstrates that Dasein's being is time;
2   a fundamental ontology, which is meant to lead to an understanding of Being *as* time;
3   and a historical destruction of the history of ontology, which allows us to retrieve the question of Being through time, i.e. through a consideration of history or tradition.

Schürmann's explication of the complex nature of this retrieval is one of the most interesting and notable features of his interpretation. Schürmann of course recognizes that most of *Being and Time* is concerned with providing an existential analytic of Dasein, and therefore with Dasein's being. However—and this is Schürmann's basic hermeneutical claim—this project can only be properly evaluated if seen in the light of the second and third projects, i.e. if seen in the light of the attempt to retrieve the question of Being as such.

This becomes clear when one takes seriously the fact that *Being and Time* is an unfinished work. In addition to an existential analytic of Dasein, *Being and Time* was to consist of a second main part deconstructing the history of ontology, which is precisely the project of Heidegger's later work. Heidegger could not finish *Being and Time* because his strategy of retrieving the question of Being by pursuing an existential analytic of Dasein could not be brought into accord with the envisaged second part of *Being and Time*. As Schürmann puts it, the basic aporia that afflicts *Being and Time* is that "in order to work out Time as the meaning or directionality (*Sinn*) of Being, Heidegger ends up working out the temporality as the meaning or directionality of Dasein" (61). Because one cannot get from the temporality of Dasein to the temporality of Being as such, Heidegger abandoned *Being and Time* and attempted to find a path of thinking that could retrieve the question of Being as such. Paradoxically, it is only by reading Heidegger "backwards" in the light of the notion of Being that became fully manifest in his later work, that one can discern not only the necessary aporia that governs *Being and Time* but also its continuity with the later work in the thesis that Being is time.

Critchley also thinks that the pragmatic interpretation of *Being and Time* obscures the importance of the question of the meaning of Being. However, he thinks that its meaning and importance is illuminated not by reading *Being and Time* through the prism of what the question means in Heidegger's later work, but by tracing the question back to its phenomenological origins. Phenomenology opens a space where this question of the meaning of Being can be retrieved because Being is in some way amenable to phenomenological seeing. As Critchley puts it, "Being is the 'seeing' of what is seen, or the 'appearing' of what appears" (9). One should not interpret this in either a subjectivist or objectivistic fashion. On the one hand, Heidegger argues that this "seeing" or "appearing" cannot be assimilated to what is subjectively given because it involves a pre-given trans-subjective correlation between an *intentio* and an *intentum*. The supposedly subjective sphere of consciousness is always already open to the world; it is, as it were, saturated by it. On the other hand, one should not assume that what is trans-subjectively given is an abstract entity, a meaning, sense, or noema.

To posit this is to interpret that which is given in the light of an objectivistic fore-understanding, which reads the conclusions of the theorist into lived experience itself. For Heidegger, Husserl's great mistake is in allowing himself to be implicitly guided by this fore-understanding. If one instead keeps to what shows itself by wresting Being from its tendency to cover itself over, one shall find a level of manifestation or disclosure that is prior to both subject and object.

While the pragmatic reading of Heidegger is thus right to say that intentionality is not the original structure of psychic life (even if it is still an important structure), this origin is not, as Dreyfus thinks, elucidated by "everyday coping skills". Rather, we must see the level of pre-given trans-subjective manifestation as the work of categorial intuition. This concept, which finds its origin in Husserl's *Logical Investigations* is, for Critchley, the essential concept that paves the way for Heidegger's path of thinking. What categorial intuition shows is that every intending is accompanied by a pre-understanding of Being. This grounds the fact that there is a pre-given correlation between an *intentio* and an *intentum*. And this, in turn, is the "basis for the early Heidegger's claim that Dasein and World must be viewed as a unitary phenomenon, and for the later Heidegger's claim for the thought of *das Ereignis* as the co-belonging or *Zusammengehörigkeit von Mensch und Sein*, the belonging together of the human being with being or that which is to be thought" (10). Categorial intuition thus not only provides the basis for the question of Being in *Being and Time*, but also for the way this question is framed and answered in Heidegger's later work. Like Schürmann, Critchley thinks that there is continuity between early and late Heidegger. This continuity, however, is phenomenological.

But if Heidegger is really just a radical phenomenologist, how is he different from Husserl? The difference with Husserl is that in overcoming an objectivistic understanding of phenomenology, Heidegger locates categorial intuition not only in intentional states narrowly construed but also in the structures of factical life itself. With this, phenomenology becomes a hermeneutics of facticity concerned not so much with the structures of a subjective consciousness but with "truth as *aletheia*...the *temporalized* bivalence of disclosure and closure, of unconcealment and concealment, or *Ereignis* and *Enteignis*" (11). In the later work, the stress shifts from a hermeneutics of Dasein's facticity to Being's temporalized unconcealment and concealment itself. However, this shift in emphasis is still underlain by categorial intuition insofar as the event of being that happens behind Dasein's back is still categorially structured and disclosed.

There is a second overall problem with the orthodox interpretation of Heidegger. While it correctly claims that Heidegger is concerned with radically rethinking the divide between theory and practice, it locates it in the wrong place. In accordance with his prioritization of Division One of *Being and Time*, Dreyfus argues that this rethinking occurs with Heidegger's demonstration that Dasein's theoretical comportment towards objectively present objects is derivative from its practical relation to ready-to-hand objects. While recognizing the importance of this reversal, Schürmann flags a more fundamental one: what Heidegger tries to show in both his early and later work is that the thought of Being is made possible by a prior practical comportment or attitude of openness. The "priority of the practical" in its deepest meaning does not concern the priority of "coping skills" but rather the institution of a "state of existence" open to Being.

In *Being and Time*, this openness—what Schürmann calls a practical a priori—is achieved in authenticity, which as anticipatory resoluteness counters Being's own tendency to cover itself up by leveling itself down into calculable and interchangeable everyday things. In authentic resoluteness, Dasein recaptures an attitude of wonder (*thaumazein*) where what is taken most for granted, i.e. Being, is defamiliarized and made strange. This is key for retrieving the question of Being, for wonder, Schürmann tells us, is precisely the "the initial attitude from which the project of *Being and Time* arose insofar as it is that attitude that is required for the retrieval of the question of Being" (116). As Schürmann notes— and this is a point that Critchley will make central to his view—the retrieval of wonder, even if covered over by the structures of falling, is a standing possibility for Dasein because at the heart of our Being there is an "a priori enigma". This enigma concerns, as Critchley puts it, "the enigma of a *Faktum*, the fact *that* one is; philosophy begins with the riddle of the completely obvious" (137).

This interpretation of authenticity is controversial, because it rejects the notion that Heidegger's disastrous political commitments can be traced back to this concept. Instead of seeing authentic resoluteness as a heroic mode of choosing ourselves in the face of death, Schürmann argues that the confrontation with death and finitude instills a mode of openness to Being that is in line with, and the origin of, Heidegger's latter concepts of letting-be, releasement, and eventually *Ereignis*. This interpretation once again flows from Schürmann's strategy of reading Heidegger backwards, which he thinks brings out the affinity of these concepts.

Critchley, in his lecture "Originary Inauthenticity—on Heidegger's *Sein und Zeit*" offers the opposite interpretation, in which Heidegger's political commitments are attributable to the complex of concepts—death, historicity, fate, and destiny—that ties the authentic resoluteness of the individual to the collective. Because—as outlined in section 74 of *Being and Time*—authenticity can only be related to the fate and destiny of a *unified* people, Heidegger cannot conceive of publicness or plurality in a way that is not denigrated as inauthentic. What Critchley wants to do is to provide a reading of *Being and Time* that vindicates this inauthentic plurality by de-emphasizing the *existentialia* that underlie the heroic reading of authenticity. "Ultimately, I would like to *modify* the way we hear the formulations, 'thrown projection' or 'factical existing', by placing the emphasis on the *thrown* and the *factical* rather than on projection and existence. That is, on my interpretation, Dasein is fundamentally a *thrown* throwing off, a *factical* existing" (138).

Critchley thinks such a reading becomes available by interpreting *Being and Time* through the prism of the previously mentioned 'a priori enigma' at the center of Dasein's being. This riddle provides the key to his reading because it concerns what is obvious and taken for granted, "the sheer facticity of what is under our noses, the everyday in all its palpable plainness and banality" (137). When one achieves authenticity, the enigmatic and uncanny quality of the everyday—and so the wonder that makes us open the Question of Being in the first place—does not disappear but instead continues to hound one's Being. Critchley is here working towards a notion of Dasein's originary inauthentic openness to Being and the other, a notion inspired by his work on Levinas. Here we find a

5

confluence in intention with Schürmann insofar as both want to find in *Being and Time* the moment of openness to Being and to the other. However, instead of finding this moment in authenticity itself, as Schürmann does, Critchley thinks we should find it in our originary inauthenticity.

* * *

All of the writings included in this book are based on lecture courses given at the New School for Social Research: Schürmann's in 1978, 1982 and 1986, and Critchley's in 2005. While Critchley's text was prepared for publication, Schürmann's lecture was never meant for publication. Three graduate students at the New School—Morgan Meis, Matthew Lear, and Steven Levine—prepared the lecture for posthumous publication. The original lecture is from the Schürmann archive located at the Fogelman Library at the New School. Although the lecture was fully written out, we have made some minor editorial changes to suit publication, such as changing first person references from the singular to the plural.

The most difficult editorial issue concerns Schürmann's idiosyncratic method of quoting *Being and Time*. While Schürmann sometimes translated directly from the German, he also used a manuscript of Joan Stambaugh's translation, as well as the Macquarrie and Robinson translation. Two difficulties must be kept in mind. First, Schürmann did not have access to the final, published version of Stambaugh's translation. Second, even when Schürmann used the Stambaugh and Macquarrie and Robinson translations, he often changed them, sometimes significantly.

We have accordingly devised a notation scheme to identify the status of each of Schürmann's references to *Being and Time*. The first reference to *Being and Time* after any quote will be to the German edition (SZ). If it reads SZ and the page number alone, this indicates that Schürmann is translating directly from the German. If there is a further notation, it means that we have been able to track down the quote to one of the English translations. If it reads JS followed by a page number, it means that Schürmann is using the Stambaugh translation, either verbatim or as a template for his own translation. If it reads MR followed by a page number, then Schürmann is using the Macquarrie and Robinson translation, either verbatim or as a template for his own translation. We have inserted italics, quotation marks, and punctuation that reflect the original German (as well as the translations) when it seems clear that Schürmann merely overlooked these or omitted them since his manuscript was never intended for publication.

We have followed MR when it comes to the question of translating the various German words for Being. There is, however, one important exception: while MR uses the word "entity" when translating "*Seiendes*", Schürmann has consistently switched back to "being". We have left "being" uncapitalized when this is the case, and have capitalized all cases where "Being" translates "*Sein*" or "*seiend*" (thus deviating from MR). This also deviates from JS, who uses uncapitalized 'Being' throughout. Thus, we have altered Schürmann's text, as well as all quotes, to fit this schema.

There are three types of footnotes used in the Schürmann lecture: Schürmann's notes, our editorial notes, and notes that include the handwritten marginal notations that Schürmann sometimes wrote out to supplement his type-written lecture. Editorial notes are in square brackets [ ], while marginal notes are in curly braces { }.

Notes

1 Dreyfus, H. (1991) *Being-in-the-World: A Commentary on Heidegger's* Being and Time, *Division One*, Cambridge, MA: MIT Press.
2 Ibid., 3.
3 Ibid.
4 It is important to note that this criticism is not specifically directed at Dreyfus. However, the position Schürmann targets, what he calls "the common thesis", would include the position Dreyfus espouses.

1

HEIDEGGER FOR BEGINNERS[1]

Simon Critchley

It's not always easy being Heideggerian.[2]
What is Philosophy? Deleuze and Guattari

Introduction

Essentially, I want to deal with two questions:

1   Where should one begin with Heidegger?
2   More importantly, why should one begin philosophizing with Heidegger rather than elsewhere?

I will try and respond to these questions by showing that the beginning of Heidegger's philosophy is *phenomenological*. That is, Heidegger's thought begins as a radicalization of Husserlian phenomenological method. The consequences of this claim for our understanding of Heidegger's work as a whole, and for the many conflicting interpretations to which it has given rise, will hopefully emerge as we proceed. To make good on my claim, I will give an interpretation of the Preliminary Part of Heidegger's important 1925 lecture course, *Prolegomena zur Geschichte des Zeitbegriffs*, a text that I see as the buried phenomenological preface to *Sein und Zeit*. Rejoining Heidegger's *magnum opus* to its phenomenological preface, permits one, in my view, to clarify the philosophical presuppositions that are required in order for *Sein und Zeit* to begin; that is, in order for the question of the meaning or truth of being to be raised as a matter of compelling philosophical interest, and not as some magical and numinous vapor.

My basic premise, to echo one of Heidegger's reported remarks from the 1962 *Protokoll* to the seminar on *Sein und Zeit*, is that "*In der Tat, wäre ohne die phänomenologische Grundhaltung die Seinsfrage nicht möglich gewesen*" ("Actually, the question of being would not have been possible without the basic phenomenological attitude").[3] If this is true, then it means that the interpretation of the *Prolegomena* assumes great importance, for it is there that Heidegger's radicalization of phenomenology is *systematically* presented as part of an

Auseinandersetzung with Husserl and not gnomically intimated, as the novice to *Sein und Zeit* often feels in reading the crucial methodological Paragraph 7 for the first time.

Heidegger's double gesture

The reading of Husserl is dominated by a double gesture that permits Heidegger both to inherit a certain understanding of Husserl, while at the same time committing an act of critical parricide against him, what von Herrmann calls the *Zweideutigkeit* or ambiguity of speaking against Husserl in Husserlian language.[4] In these lectures, I would like to sketch this double gesture in some detail.

For Heidegger, there are three essential discoveries of Husserlian phenomenology: intentionality, categorial intuition, and the original sense of the a priori. These discoveries are linked together in a "nesting effect", where intentionality finds what Heidegger calls its "concretion" in categorial intuition, whose concretion is the a priori, which provides, in turn, the basis for a new definition of the *Vor-Begriff*, the preliminary concept of phenomenology itself, a definition that is only accidentally modified in Paragraph 7 of *Sein und Zeit*. I believe that this definition of phenomenology remains at least formally determinative for the rest of Heidegger's philosophical itinerary. To put this into a schema:

intentionality + categorial intuition + the a priori = the preliminary concept of phenomenology

I shall elaborate these concepts in more detail presently, but it should be noted that the condition of possibility for Heidegger's concept of phenomenology is a certain understanding of the intentionality thesis. However, as Heidegger puts it, although intentionality is the essential structure of mental experience—what Heidegger calls 'psychic life'—it is not the original structure, which is given in the analysis of categorial intuition.

As we all know, Heidegger's thinking is preoccupied—perhaps a little too preoccupied, but that is another story for a separate occasion—with the *Seinsfrage*, the question of being. Phenomenology opens a space where the question of being can be raised, releasing being from the subjectivistic determination to which it had been submitted in philosophical modernity, most obviously in Descartes, Kant, Fichte and others, but more closely in the Neo-Kantianism of Heidegger's peers and superiors in Marburg.

Heidegger's leading, but hardly self-evident, philosophical claim, which I shall try to clarify below, is that being is an aspect of phenomenological seeing, in some sense a matter for phenomenological intuition. We might say that being is the 'seeing' of what is seen, or the 'appearing' of what appears, although this should not be misunderstood as announcing some sort of metaphysical dualism. Thus, against the modern philosophical self-understanding, phenomenology grants to being a new sense of non- or, better, *trans*-subjective givenness. As

Klaus Held insightfully remarks, Husserl's discovery for Heidegger is "*die Vorgegebenheit einer transsubjectiven Offenbarkeitsdimension*" ("the pre-givenness of a trans-subjective dimension of manifestation").[5]

As the work of Jacques Taminiaux has shown in detail, the pre-givenness of this trans-subjective dimension of manifestation is the work of categorial intuition.[6] When Heidegger famously remarks at the end of Paragraph 7 of *Sein und Zeit* that the latter book only became possible "*auf dem Boden*" laid down by Husserl, then this *Boden*, this ground or basis, alludes to categorial intuition (SZ 38). The central position that Heidegger gives to categorial intuition in the interpretation of Husserl and to Heidegger's self-understanding as a phenomenologist remains unaltered from *Sein und Zeit* to the final seminar in Zähringen in 1973. In this sense, we might say that Heidegger's real contribution to philosophy is his radicalization of the basic idea of phenomenology, a radicalization that paradoxically shows the extent of his debt to Husserl, and, by extension, the radicality of Husserlian phenomenology.

As Heidegger points out in 1963, with an explicit look back over his shoulder to the very same lines from Paragraph 7 of *Sein und Zeit* that were cited above, phenomenology must not be understood as a movement or school, but as the possibility of thinking as such. That is, phenomenology is the possibility of corresponding to the claim of that which is to be thought ("...*dem Anspruch des zu denkenden zu entsprechen*").[7] For the early Heidegger, what is to be thought is the meaning of being, and for the later Heidegger, the truth of being. In these lectures and my interpretation of *Sein und Zeit*, we will hopefully begin to understand what he means by the meaning or truth of being. In my view, the latter is intrinsically bound up with what I call below the "openedness" of thrown projective finite Dasein, but this will hopefully become clearer below.

Importantly, in order to conceive of the task of thinking as a correspondence between thought and that which is to be thought, what has to be presupposed is the idea of phenomenological correlation that Heidegger finds in the intentionality thesis and pursues in his analysis of categorial intuition. It is this idea of a phenomenological correlation irreducible to either subjectivism or objectivism that is the basis for the early Heidegger's claim that Dasein and World must be viewed as a unitary phenomenon, and for the later Heidegger's claim for the thought of *das Ereignis* as the co-belonging or *Zusammengehörigkeit von Mensch und Sein*, the belonging together of the human being with being or that which is to be thought. The thought of phenomenological correlation thus bridges any idea of a "Heidegger 1" and a "Heidegger 2" and problematizes the whole idea of the *Kehre*, or a turn in thinking that is alleged to take place in the mid-1930s when Heidegger is meant to move from the human being to being as such.

The classic statement of this view lies in Richardson's hugely impressive 1962 text, *From Phenomenology to Thought*.[8] In my view, the unity of Heidegger's work is phenomenological. His difference with Husserl is that the thought of phenomenological correlation is deepened, firstly, by the claim, inherited from Dilthey, into the primacy of factical life that requires a corresponding mode of practical or hermeneutic insight; and, secondly, by the claim for truth as *aletheia*, as the *temporalized* bivalence of disclosure and closure, of unconcealment and concealment, or *Ereignis* and *Enteignis*.[9] But this will become clearer as we proceed.

I would like to try to reconstruct the various conceptual moves that permit Heidegger to deduce his *Vor-Begriff* of phenomenology. I will pass over much of the fascinating intellectual history of these concepts, which is very clearly laid out by Heidegger in the opening pages of the *Prolegomena*, and try to bring out the central arguments that will hopefully philosophically justify those concepts. Let me begin with the intentionality thesis.

Intentionality

For Heidegger, intentionality is the essential structure of subjectivity *qua* Dasein. Mental experience is fundamentally characterized by what Heidegger calls *Sich-richten-auf*, directing itself towards. That is, mental experience is always directing itself towards its matters, it is always already outside, alongside and amidst things and not enclosed in what Heidegger calls "the cabinet of consciousness". This understanding of the intentionality thesis can be found in a passage from Paragraph 13 of *Sein und Zeit* that is much less dramatic when read in the English translation. Heidegger writes:

> *Im Sichrichten auf...und Erfassen geht das Dasein nicht etwa erst aus seiner Innensphäre hinaus, in die es zunächst verkapselt ist, sondern es ist seiner primären Seinsart nach immer schon "draußen" bei einem begegnenden Seienden der je schon entdeckten Welt* (SZ 62).

As Heidegger is reported to have said in his final seminar at Zähringen in 1973, *Dasein ist das Ekstatische*. That is, the fundamental quality of mental experience is not found in the immanence of consciousness, but is rather *Da*, it is had there, outside, alongside things, and not divorced from them in a mental capsule full of representations. Put another way, mental experience is fundamentally transcendent. As Heidegger, alluding to Kant, puts it in *The Basic Problems of Phenomenology*, intentionality is the *ratio cognoscendi* of transcendence, and transcendence is the *ratio essendi* of intentionality.[10]

The intentionality thesis permits Heidegger to make the passage from *Bewußtsein* (consciousness) to Dasein in a reading of Husserl which, beneath the apparent generosity, ultimately works against Husserl's intentions. That is, beneath the surface of the exposition, Heidegger has already insinuated an anti-Husserlian pre-theoretical model of intentionality, what we might call a *phronetic* or practical intentionality or, following Taminiaux, "the Aristotelianization of Husserl".[11]

But let's look at this claim a little more closely. Heidegger argues that natural perception, of the most everyday kind, is not a detached theoretical observation, but is rather completely absorbed in our dealings with things. That is to say, perception is not self-contained, it is not contained in a self that is divorced from what it sees, but is rather already contained in a world that is familiar to us and constitutive of "self". Of course, against this simple and quite naive (I shall come back to naiveté) account of perception, one can rightly object that it introduces the unargued metaphysical assumption that the mental comes out of itself to correspond to the physical, that consciousness is in immediate contact with reality. A naive realist version of intentionality can be refuted by showing how natural perception is continually subject to the possibility of deceptive perception or hallucination, e.g. Descartes' example of a straight stick appearing bent when placed in a river, or Heidegger's more surrealistic example of imagining an automobile being driven through the lecture theatre over the heads of his auditors (PGZ 38/HCT 30). If this objection is correct, then a better interpretation of natural perception would be to claim that I do not perceive objects themselves, but rather only the psychic contents that the impressions of those objects leave on my consciousness—a representationalist theory of perception. The evidence of deceptive perception would seem to lead to the inevitable conclusion that a distinction should be made between subjective consciousness and its objects, which is precisely the position Heidegger wants to refute.

But does this interpretation of deceptive perception in any way refute the properly phenomenological concept of intentionality? "*Nein*", says Heidegger very firmly (PGZ 39/HCT 31). He turns this argument on its head by firstly assenting to the evidence of hallucination, and agreeing that in the case of deceptive perception there is no correspondence between the mental and the physical, but secondly by going on to insist that a hallucination, although a deceptive perception, is still a perception *of* something, and is still therefore intrinsically intentional. Recalling Brentano's thesis of intentional inexistence as reinterpreted by Husserl in the fifth *Logical Investigation*, Heidegger insists that perception is still intentional even in the absence of a real object, and furthermore that it is precisely in virtue of the fundamental intentionality of perception that such deceptive perception or hallucination becomes possible. Heidegger concludes that "When all epistemological assumptions are set aside, it becomes clear that comportment itself [*die Verhaltung selbst*]...is in its very structure a directing-itself-towards [*Sich-richten-auf*]" (PGZ 40/HCT 31).

Having established this first empty specification of intentionality, and introduced Husserl's notion of the intentional object as something irreducible to either a real object or a subjective impression, Heidegger goes on to describe the content of the basic constitution of intentionality. He does this by addressing three foci:

- first, the perceived thing in perception, the thing or entity in itself (*das Wahrgenommene des Wahrnehmens, das Seiende an ihm selbst*);
- second, perceiving or perceivedness as such (*Wahrgenommenheit*), i.e. the different ways in which an entity is intended;
- third, the belonging together (*Zusammengehörigkeit*) of the perceiving and the perceived.

With regard to the first focus, if intentionality is understood as directing-itself-towards, then towards what is intentionality directed? What is perceived in perception? Taking the example of a chair, Heidegger responds that I see the chair. That is, I do not see a representation of the chair, but rather the chair itself standing before me in the lecture theatre as an article available and ready for my use. That is, first of all I perceive the chair as a thing in my environment (*Unweltding*, which recalls Aristotle's notion of *ta techne onta*), on to which can be grafted a perception of the chair as a natural thing (*Naturding, ta physei onta* in Aristotle). Heidegger's point is that my perception of the chair as a natural thing only begins when I no longer regard it as an environmental item available for my use, but rather examine the chair contemplatively, assessing its various qualities, its weight, height, width, color and the wood from which it is made. That is—and I shall come back to this at length later on in the lectures—to perceive things as natural things is to adopt the theoretical and theoreticist attitude of philosophy and the sciences. But the essential point here is that the perceived thing is experienced *both* as a natural thing *and* as an environmental thing, for example I can engage in a little speculative carpentry, or I can just sit in the chair. In both cases, it is the same chair that is perceived; what changes is the way in which I see it, where I engage in what might be called categorial aspect-change with respect to the thing.

Heidegger illustrates his point with the example of two experiences of a rose—an example with a singular destiny in Heidegger's work, as is shown by the discussion of Angelus Silesius in *Der Satz vom Grund*—first as a flower that might happen to grow in my garden and that I might decide to give to someone, and second as a plant that I might subject to a botanical analysis.[12] The rose as a flower in the practice of gift-giving functions as an environmental thing, and it would be somewhat infelicitous to say, "Here, my dear, are some flowers from the shrub or climbing plant of the rosaceous genus, which quite typically have prickly stems and compound leaves". On the other hand, it would be equally inappropriate for a botanist charged with the complex taxonomy of various new rose hybrids, to simply quote Coverdale's lines, "As the rose among the thorns, so is my love among the daughters".

However, although Heidegger emphasizes how the same thing can be perceived in two ways, he is also anxious to establish the relation of the priority between the natural and the environmental experience of things, or the *vorhanden* and the *zuhanden*, the present-at-hand and the ready-to-hand. As he shows in *Sein und Zeit*, the experience of things as natural emerges out of, and is founded upon, a prior environmental understanding of the thing and the world of which it forms a part. Thus, the theoretical perception of things is secondary to, and derived from, the practical and factical experience of an inhabited environment that is familiar and common. As Heidegger puts it in *Sein und Zeit*: "Knowing is a mode of Dasein founded upon Being-in-the-world" (SZ 62).

But, it is important to point out, the move from a theoretical to a pre-theoretical comportment towards things is not a move to a pre-intentional level. Rather, what Heidegger tries to argue for—*inter alia*—in *Sein und Zeit* is an affective or

14

mood-ful intentionality, an intentionality of *Stimmung*. As Heidegger remarks in Paragraph 29 of *Sein und Zeit*, "*Die Stimmung hat je schon das In-der-Welt-sein als Ganzes erschlossen und macht ein Sichrichten auf...allererst möglich*" ("Mood has already disclosed, in every case, being-in-the-world as a whole, and makes it possible first of all to direct oneself towards something", SZ 137).

If someone, say one of Heidegger's Neo-Kantian colleagues at Marburg, were to object that this position is unscientific and naive, then Heidegger would respond, "In opposition to this scientific account, we want precisely naiveté, pure naiveté, which in the first instance and actuality sees the chair" (PGZ 51/HCT 39). Seeing here does not mean optical sensing or theoretical contemplation, it rather describes the naiveté of simply taking stock of what is found. And although Heidegger explicitly insists that such seeing does not entail "a mystical act or inspiration" (PGZ 52/HCT 40), one might suggest that this experience of naive seeing, the simplicity of this pre-theoretical regard, is the very facticity of the mystical in Angleus Silesius's experience of the "rose without why".

After having considered the thing perceived, Heidegger turns to the second focus, the perceiving of the thing. This moment has priority in the analysis insofar as any phenomenology of perception is not primarily concerned with the perceived thing, but rather with the different intentional modes of the thing's perceivedness (*Wahrgenommenheit*). Thus what interests Heidegger are the manifold ways in which things are intended, what he calls "the how of being-intended" (*das Wie des Intendiertseins*). Heidegger's taxonomy identifies four modes of intentional relating to things.

1   Bodily given: things can be perceived in terms of sheer bodily presence (*Leibhaftigkeit*). For example, the chair in front of me is perceived as being bodily there (*Leibhaft-da*), where I am simply there with the thing that I perceive.

2   Self-given: in addition to relating to a thing as something bodily given, I can—and this distinction will prove crucial in the discussion of categorial intuition—intend it as self-given (*Selbst-gegeben*), where I envisage a thing that is not bodily there like the chair. This is illustrated with an example:

I can now envisage the Weidenhauser bridge: I place myself before it, as it were. Thus, the bridge is itself given [*selbst gegeben*]. I intend the bridge itself and not an image of it, no fantasy, but it itself. And yet it is not bodily given [*Leibhaft gegeben*] to me. It would be bodily given if I go down the hill and place myself before the bridge itself. This means that what is itself given [*selbst gegeben*] need not be bodily given, while conversely anything which is bodily given is itself given.

(PGZ 54/HCT 41)

3   Empty intending: the character of self-givenness, which is a non-sensuous but intentional intuiting of things, can be more clearly delineated by contrast with what Heidegger calls "empty intending" (*Leermeinen*). This is a way of intending a thing without really intuiting it in the manner of self-givenness. For example, the project of renovating or replacing the Weidenhauser bridge might crop up as part of a conversation, and I would emptily intend the bridge without actually seeking the intuitive fulfillment for such an intention that takes place in the modes of self-givenness and bodily-givenness. Empty intending predicts much that Heidegger tries to catch with the notion of *Gerede*, or idle talk, in *Sein und Zeit*: "A large part of our ordinary talk [*natürlichen Rede*] goes on in this way. We mean the matters themselves and not images or representations of them, yet we do not have them intuitively given" (PGZ 54/HCT 41).

4   Heidegger distinguishes a fourth intentional mode with what he calls "picture perception" (*Bildwahrnehmung*), where, for example, I can be intentionally related to the Weidenhauser bridge by looking at a photograph or picture postcard of it. In this mode of intending, what is bodily given is the picture postcard. Heidegger's reasons for introducing this final mode of intending are less than clear, except that he seems to be offering an implicit critique of any representationalist theory of perception, where my relation to the world is mediated through the various pictures that sense data leave on consciousness. For Heidegger, like Hegel, experience cannot be reduced to picture thinking (see SZ 214–15).

Heidegger insists that the authentic moment (*das eigentliche Moment*) of perception takes place when the perceived is bodily given, and that these four modes of intending things are structurally related to each other in a hierarchy where a lower level of intending can be fulfilled by a higher level (PGZ 43/HCT 57). Empty intending is fulfilled by self-givenness, which is in turn fulfilled by picture perception, which attains complete fulfillment in bodily givenness.

Moving on to the third focus, once the different modes of intending have been delineated, then the basic constitution of intentionality can be brought into view. This is the belonging together of the *intentio* and the *intentum*, to borrow the somewhat artificial Latinate terminology employed by Heidegger to avoid Husserl's talk of consciousness and its objects (PGZ 58–63/HCT 44–47). In many ways, all that Heidegger is trying to establish in his account of intentionality is the intimate affinity or identity between the *intentio* and the *intentum*, a belonging together that demonstrates the basic structure of mental experience as a directing-itself-towards, a movement of transcendence or ecstatic openness.

What Husserl's intentionality thesis gives to Heidegger is a basic structure that begins from the phenomenological correlation of the human being to things in a way irreducible to either subjectivism or objectivism. However, within this seemingly generous and uncritical reading of Husserl—Heidegger even goes so far as to assimilate the *intentio/intentum* distinction to a non-idealistic reconstruction of the *noesis/noema* distinction (PGZ 60–61/HCT 45–46)—Heidegger is already demanding "a more radical internal development of phenomenology".

At this early stage of the analysis, it is already clear that, for Heidegger, it is not intentionality that is metaphysically dogmatic, but rather "what is built under its structure" (PGZ 63/HCT 47), that is the presupposition that mental experience can be determined in advance as consciousness. It is this dogmatism, of what Heidegger rather euphemistically calls "a traditional tendency", whose source is Descartes, that will eventually overtake the radicality of Husserlian phenomenology. For Heidegger, such a tendency has no place in phenomenology and a more radical internal development of Husserl must—and note that Heidegger is already alluding to the definition of phenomenology given in Paragraph 7 of *Sein und Zeit*—keep to *was sich selbst zeigt*, that which shows itself (PGZ 63/HCT 47). For Heidegger, radicality means keeping to that which shows itself, and, furthermore, that which shows itself most closely and mostly, as Heidegger repeats endlessly in *Sein und Zeit*, *zunächst und zumeist*, proximally and for the most part.

As I said above, intentionality is the essential structure of subjective life for Heidegger, but it is not the original structure. To clarify this original structure, Heidegger insists that it is necessary to follow intentionality into what he calls its "concretion", which is only given with the concept of categorial intuition.

Categorial intuition

No one can tell, of the things he now holds in his hand and reads, how much comes in through his eyes and fingers and how much from his apperceiving intellect, unites with that and makes of it this particular "book"? The universal and the particular parts of the experience are literally immersed in each other, and both are indispensable. Conception is not like a painted hook, on which no real chain can be hung; for we hang concepts upon percepts, and percepts upon concepts interchangeably and indefinitely; and the relation of the two is much more like what we find in those cylindrical "panoramas" in which a painted background continues a real foreground so cunningly that one fails to detect the joint. The world we practically live in is one in which it is impossible, except by theoretic retrospection, to disentangle the contributions of intellect from those of sense. They are wrapt and rolled together as a gunshot in the mountains is wrapt and rolled in fold of echo and reverberative clamor. Even so do intellectual reverberations enlarge and prolong perceptual experience which they envelop, associating it with remoter parts of existence.

William James[13]

I would like to approach what Heidegger called in 1973 "the burning point" (*Brennpunkt*) of Husserl's thought in three steps. It will take us the next two lectures to get through this material. First, I will explain the huge significance of categorial intuition for Heidegger by explaining its relation to both a polemical understanding of the categorial in modern philosophy and, against that modern understanding, the attempted phenomenological retrieval of the ancient ontology of Plato and Aristotle. Second, I will give a summary of Heidegger's treatment of categorial intuition in the *Prolegomena*, as this will enable a number of essential distinctions to be made: between perception and assertion, simple and complex acts, and acts of synthesis and acts of ideation. Third, I will step back from the textual and historical thickets of Heidegger and attempt to give a more distant rational reconstruction of the concept.

I

What are categories? Categories are what we might call "meta-concepts", which are required in order to explain the way in which human beings understand the things presented in experience. That is, categories are required in order to explain the way in which the perceptual experience of things is *de facto* conceptually articulated or made intelligible. As such, categories are a *de jure* distinguishing device for understanding the use of concepts that is based upon the pre-existing *de facto* unity of perception and conception, so eloquently described above by William James.

For example, I may perceive a chair, and say the word "chair" while pointing at the four-legged wooden thing before me. Such would be the raw translation of a sensuous percept into a simple concept. However, to go on and utter the assertion, "this is a chair", through which my perception becomes more articulated and independently communicable, introduces the copula, the third person present indicative of the verb to be, and consequently the category of substance or being.

Indeed, this is something that can be observed in childhood language acquisition, when the infant moves from the simple name, for example "ball"—a name that is habitually and amusingly (for the parents) extended to many non-ball objects and where it will describe, say, anything circular, such as the moon—to the assertion "it's a ball". Again, if I say "the chair is brown", then this means that I perceive the chair and see brownness as a quality of the chair, but these perceptual features are articulated through the insertion of the copula. And the purpose of such articulation is the reduction of possible ambiguity. For example, imagine the connotative ambiguities of "chair brown" or "ball blue" without the mediation of the copula.

Now, the whole discussion of categorial intuition turns on the status of the copula: what is the "is"? In the assertion, "this is a chair", I see the chair, I see its color, its composition and construction. I can feel the texture of the wood from which it is made and—in the intentional mode of self-givenness—I can even present to myself the forest or wood from which the tree was taken, how the tree was felled and turned into timber to be transported to a workshop or furniture factory. That is to say, I can find or imagine intuitive fulfillment for all sensuous aspects of the chair, but in none of this do I see the "is". In short, I can find no sensuous fulfillment or intuitive evidence for being. What, therefore, is the status of such non-sensuous categorial forms? What evidence or possible intuitive fulfillment is available for the category of being? Is being something seen in a non-sensuous manner? If so, then how?

18

The categories were first introduced by Aristotle in the *Topics* as those characteristics that can be predicated in things, for example substance ("the chair *is* brown"), quality ("the chair is *brown*"), quantity ("*the* chair is brown"), relation, place, time, position, state, action and passion.[14] It is this Aristotelian conception of the categories that is alluded to in the enormously suggestive passage from Paragraph 9 of *Sein und Zeit*, where Heidegger distinguishes categories from existentials, that is, items that aim at a "what" (*ein Was*) from items that aim at a "who" (*ein Wer*), namely a Dasein that exists.[15] Heidegger writes: "The *kategoriai* are what is sighted and what is visible in such a seeing" ("*Das in solchem Gesichtete und Sichtbare sind die kategoriai*") (SZ 45). Furthermore, the categories are—and we shall return to this formulation below—what permit a *sehen lassen*, a letting be seen, which is Heidegger's rendering of the notion of *logos* in Aristotle, a letting be seen or *be-speaking* (*Besprechen*) of things that are common or public (*für Alle sehen lassen*).

For Kant, the categories are the a priori concepts that provide the conditions of possibility for understanding and unifying the manifold of intuition, or the plurality of perceived items. They are the conditions of possibility for all cognition, that is, of the placing of intuitions under concepts that take place in the de facto synthesizing activity of human cognition.[16] Now, crudely stated, the notion of categorial intuition is the thought that the being of things, that in virtue of which things are intelligible or understandable, their substance, quality or whatever, is something intuited, sighted or seen in those things; it is something intentionally given.

As will become clear in Heidegger's critique of Husserl, which I will talk about in a later lecture, Husserlian phenomenology falls fatally prey to the massive privileging of *Vorhandenheit*, that is, theoreticism or intellectualism, where the intentional comportment towards things is conceived as a relation to *objects* that stand over against a subject and which are seen in a constant presence (SZ 96). Yet, despite this, Husserlian categorial intuition allows Heidegger to approach the categorial as something sighted or visible in the things themselves.

In terms of the modes of intentional relatedness discussed above, categorial intuition has the character of self-givenness, a non-sensuous but intuitively fulfilled intuition. Thus, the doctrine of categorial intuition permits Heidegger to contest the Kantian account of the categories as the pure concepts of the understanding, where the mere suggestion of something like an intuition of the categorial would be a contradiction in terms, or would be dismissed as a return to some pre-modern notion of intellectual intuition.[17] By definition, the categories cannot be intuited for Kant because they are that in virtue of which intuitions are placed under concepts, thereby guaranteeing knowledge. As Heidegger makes clear in the *Zähringen* seminar, the Kantian categories are a function of the understanding and judgment, and therefore, to twist Gadamer's words slightly, one might speak of *the subjectivization of the categorial* in Kant.[18]

Of course, this is something of a caricature of Kant—a Cartesian Kant—as Heidegger himself came to realize in the detailed reading of Kant that followed the *Prolegomena*, which can be seen by Heidegger's extremely balanced judgment on the question of the subject and the schematism in Paragraphs 6 and 64 of *Sein und Zeit*, and which was extended in *Kant and the Problem of Metaphysics* in 1929. However, the critical claim is justified insofar as Kantian epistemology is dominated by the opposition between sensibility and understanding, where intuition is confined to the former and concepts and categories are restricted to the latter.

This point can be developed by turning to Husserl's critique of Kant towards the end of the sixth *Logical Investigation*, where he claims that the shortcoming of Kantian epistemology consists in its failure to extend the concepts of perception and intuition into the categorial realm, and that consequently Kant fails to grasp the character of pure ideation, or what he calls "an intuition of the universal". Husserl claims that "Kant drops from the outset into the channel of a metaphysical epistemology".[19] Therefore, for Husserl as well as Heidegger, there is a subjectivization of the categorial in Kant. But let us note the strategy that is at work in Heidegger: he works against Kant—or a straw man version of him—and against the understanding of the categorial in modern philosophy, and uses Husserl as an accomplice for a retrieval of a non-subjectivistic Aristotelian doctrine of the categories as the characteristics sighted in things. Thus, the discussion of categorial intuition permits one to understand how Heidegger could both claim to be a continual learner in relation to Husserl, and how Aristotle was the first true phenomenologist.[20]

The direction of Heidegger's analysis of categorial intuition is similar to that of the discussion of intentionality. Once again, it is a question of a seemingly generous, at times word for word, reading of Husserl, that eventually takes Husserl precisely where he does not want to be taken. If the discussion of intentionality follows the passage from a conception of the human being qua consciousness and subjectivity to the ecstatic openness of Dasein as being-in-the-world, then categorial intuition is the deepening of intentionality by exhibiting its a priori structure. The concept of categorial intuition permits Heidegger to show that the categorial forms employed in assertions are not unfulfilled constructions or functions of the understanding. Rather, Heidegger writes:

> The categorial "forms" are not constructs of acts but objects which manifest themselves [*sichtbar werden*] in these acts. They are not something made by the subject and even less something added to the real objects, such that the real entity is itself modified by this forming. Rather they actually present the entity more truly in its "being-in-itself".
>
> (PGZ 96/HCT 70)

The non-sensuous categorial forms of thought are not the constructs of mental acts but rather the intentional objects of those acts that "constitute a new objectivity" (PGZ 96/HCT 71). Constitution does not mean constructing something in the sense of fabrication or fabulation, but rather—and this once again pre-empts Heidegger's *Vor-Begriff* of phenomenology—"letting the entity be seen in its objectivity" (*Sehenlassen des Seienden in seiner Gegenständlichkeit*). What is clearly being opposed here is "the subjectivization of the categorial", a prejudice whose ancestry Husserl traces back to Locke and which can be traced forward to Viennese logical positivism, in Carnap, say, where "being" is dismissed as a fiction and replaced with an existential quantifier. For the empiricist tradition, the origin of the concept of being and other categorial forms "arises through *reflection upon certain mental acts, and so fall in the sphere of 'inner sense', of 'inner perception'*",[21] that is, being is an empty and ambiguous product of inward reflection.

Now, Husserl accepts the Kantian thesis that being is not a real predicate, that is to say that being is not a real constituent of the object, in the sense of something perceptible or having substantial fulfillment, but he wants to refuse the subjectivistic consequences of the Kantian thesis, namely that the origin of the concept of being can be traced to reflection, inward perception or a judgment of the understanding. In the sixth *Logical Investigation*, Husserl writes:

> The thought of a judgment fulfills itself in the inner intuition of an actual judgment, but the thought of an "is" does not fulfill itself in this manner. Being is not a judgment nor a constituent of a judgment. Being is as little a real constituent of some inner object as it is of some outer object, and so not of a judgment. In the judgment—the predicative statement—"is" functions as a moment of meaning [*Bedeutungsmoment*], just as perhaps, although otherwise placed and functioning, "gold" and "yellow" do. The *is* itself does not enter into the judgment, it is only meant, signitively referred to, by the little word "is". It is, however, self-given [*selbst gegeben*], or at least putatively given, in the *fulfillment* which at times invests the judgment, the *becoming aware* of the state of affairs supposed.[22]

Please listen carefully: the copula does not self-referentially reside in a subjectively determined judgment. Rather, insofar as it functions in a proposition, the copula is a *Bedeutungsmoment* that refers to something outside judgment in a similar way to the sensuous elements in a proposition, like "gold" or "yellow". Thus, if we return to our example "the chair is brown", then Husserl's claim is that this use of the copula and other categorial forms finds fulfillment in a way that is *analogous* (and we shall explore the nature of this analogy presently) to the way in which the sensuous elements in the proposition find intuitive fulfillment. Thus, although Husserl grants that categorial forms like being can only be apprehended through judgment, through the discursive articulation of what Heidegger would call *logos*, this does not entail "...that the concept of being must be arrived at 'through reflection' on certain judgments, or that it can ever be arrived at in this fashion".[23] For Husserl, there is a *Sachverhalt*—a state of affairs, or a matter towards which I comport myself—that invests the judgment, and which provides the intuitive resources for the sensuous and categorial fulfillment of the judgment.[24] Elsewhere, Husserl writes, "In the judgment a state of affairs 'appears' before us, or, put more plainly, becomes intentionally objective to us".[25] Husserl continues, in a famous quote, itself cited by Heidegger (PGZ 79/HCT 59):

Not in **reflection** *upon judgments, nor even upon fulfillments of judgments, but in the* **fulfillments of judgments themselves [Urteilserfüllungen selbst]** *lies the true source of the concepts of State of Affairs and Being* [in the copulative sense]. Not in these *acts as objects*, but in *the objects of these acts*, do we have the abstractive basis which enables us to realize the concepts in question.[26]

Thus, it is not in empty mental acts of reflection upon judgments, nor even in reflection upon the fulfillments of judgments, but rather *in* the fulfillments of judgments themselves, in the states of affairs or intentional objects that give meaning to those acts, that the origin of the concept of being can be found. There is an irreducible *Selbstgegebenheit* of being, or possible intuitiveness of the categorial, which means, crucially for Heidegger, that the question of the meaning of Being can be raised. The concept of intentionality and the investigation of categorial intuition as the a priori concretion of intentionality, gives Heidegger the *Boden* of the self-givenness of being that is the formal-methodological condition of possibility for raising the question of the meaning of being as such, and transforming our vague and average *Seinsverständnis* into a fundamental ontology. In other words, the movement from the first to the second discovery of phenomenology, from intentionality to categorial intuition, perhaps allows us to understand why Heidegger should choose an existential analytic of Dasein as his preliminary way to the question of the meaning of Being.

The importance of the notion of categorial intuition for the project of fundamental ontology cannot be overstated, but the strategy is complex and needs to be delineated in a number of steps:

1   Heidegger interprets the notion of categorial intuition to show that the concept of being has a trans-subjective givenness for Husserl. This means that there is:

2   an evidence or truth to the concept of being, which allows Heidegger to interpret the task of phenomenology as being directed towards the question of being; the latter becomes a matter for phenomenological seeing. Of course, this means that:

3   the true vocation of phenomenology is ontology, science of being, or metaphysics in the Aristotelian sense. And vice versa that any ontology must be phenomenological in character, "There is no ontology *alongside* a phenomenology. Rather, *scientific ontology is nothing but phenomenology*" (PGZ 98/HCT 72). If this is true, then:

4   Heidegger has located the point where phenomenology crosses the path of ancient ontology in the work of Plato and Aristotle. Now, for Heidegger, this retrieval of ancient ontology is essential because of what he perceived to be the limitations of Husserlian phenomenology. That is, although Husserl grasped the givenness of being as a matter for phenomenological inquiry, he failed to pursue this discovery with sufficient radicality, and was content to

22

determine being, to quote Heidegger's Zähringen seminar, as object-being (*Gegenstand-Sein*).[27] Where Heidegger steps beyond his teacher is in his (re)discovery—achieved, as he puts it, "from thinking through Plato and Aristotle in a questioning manner"—that objectivity is a mode of presencing (*Anwesenheit, parousia*, SZ 25). Of course, if the meaning of being in ancient ontology is presencing, then this reawakens the link that connects being with *time*, where a radicalized interpretation of temporality is the horizon for any possible understanding of being. But that, as they say, is another story.

Let's go back a little and ask: is the notion of categorial intuition convincing? To explore this, permit me a final quote from the sixth *Logical Investigation*, which concludes the run of argumentation that I have been tracking in Husserl and expresses the full concept of categorial intuition.

> It is in fact obvious from the start that, just as any other concept (or Idea, Specific Unity) can only "arise", i.e. become *self-given* to us, if based on an act which sets some individual instance of it before our eyes, so the concept of Being can only arise when *some being, actual or imaginary, is set before our eyes* [so kann der Begriff des Seins nur entspringen, wenn uns *irgendein Sein, wirklich oder imaginativ, vor Augen gestellt wird*]. If "being" is taken to mean predicative being, some *state of affairs* must be given to us, and this by way of an *act which gives it, the analogue* [das Analogon] *of common sensuous perception*.[28]

For Husserl, the concept of being does not arise merely from a blind intention, reflection or subjective judgment, rather it only arises when being itself is some-how given to us, placed before our eyes as a state of affairs in a way that is analogous to sensuous perception.

Many skeptical questions can be raised here: for example, what would the "actual or imaginary" seeing of *some being* mean here? What are the limits of the actual and the imaginary? On Husserl's account, what intentions would *not* be capable of intuitive fulfillment? Would the notion of categorial intuition entail that it is meaningful to speak of anything, such as God or the gods, Hades and the world of the Hobbit? Also, what is the precise nature of the *analogy* between sen-suous and categorial intuition? It is clear that Husserl arrives at the concept of categorial intuition through an argument from analogy, where sensuous intuition provides the standard for categorial intuition and where the latter may be said to be analogically derived from the correlative structure of the former. But how is this argument from analogy consistent with the claim that although the sensuous is the founding mode for the categorial, which is a founded mode, the categorial has priority in the analysis, and indeed provides the a priori condition of possibil-ity for sensuous intuition?

Cobb-Stevens has claimed that this complication of the founding/founded dis-tinction in sensuous and categorial intuition is "a productive paradox"; perhaps that's true, but the question has to be answered as to why this is not simply a confusion or a vicious paradox.[29] As Heidegger points out in the Zähringen seminar, does not the concept of categorial intuition seem to demand some sort of *double seeing*, both sensuous and categorial; and doesn't this return us to certain well known metaphysical ambiguities as ancient as Plato's theory of forms, where the seeing of the *eidos* is also claimed to be analogous with phenomenal seeing, but where one might object, with the cynic Antisthenes, "Plato, I see the horse, but I do not see horseness"?

I do not think that such questions amount to a wholesale dismissal of categorial intuition, but rather necessitate a rational reconstruction of the concept. I will turn to this presently, but first certain essential distinctions will have to be made by returning more closely to Heidegger's *Prolegomena*.

II

Heidegger approaches the doctrine of categorial intuition through an interesting, but rather loosely argued, discussion of the assertion or proposition (*Aussage*). Assertions are acts of meaning; that is, they are sentences that express or speak out (*ausdrücken, aussagen*) perceptual experience. As such, our concrete perceptions are utterly pervaded by the assertions that articulate them, much in the manner evoked above by William James in the arresting, and very American, image of a gunshot reverberating in the surrounding mountains. For Heidegger, and this is arguably his difference with Husserl, it is highly misleading to speak of the priority or antecedence of perception over expression, or of intuition over concept, as if one first looked at a thing and then, and only then, articulated this perception in an assertion. If anything, the order of priority should be reversed, and Heidegger suggests, in an anticipation of the disclosive function of *Rede* in *Sein und Zeit*, that we first see things when we talk about them, "...we do not say what we see, but rather the reverse, we see what one says about the matter" ("*sprechen wir nicht das aus, was wir sehen, sondern umgekehrt, wir sehen, was Man über die Sache spricht*") (PGZ 75/HCT 56).

Although he does not develop the issue, I do not think that Heidegger is denying the *fact* of pre-linguistic perception, in infants say, rather he is suggesting that such perception only becomes *meaningful* insofar as it is mediated through language. The world is first meaningfully disclosed in words, and our perceptual seeing of things is structured by a conceptual saying that speaks out or expresses those things. If, as both Husserl and Heidegger insist, there is an "objectivity" to the categorial forms employed in assertions, then this means that the intuitive self-givenness of being is a seeing that occurs through a saying. As the later Heidegger will remark, language is the house of being. Being is given in language and, as Heidegger says, alluding to Stefan George, "where word breaks off no thing may be".

However, Heidegger's guiding question in these pages is already familiar: "How can we call an assertion true when we make it within a concrete perception? Can the assertion which I make in a concrete and actual perception be fulfilled in the same way that an empty intention corresponding to the concrete perception is fulfilled?" (PGZ 75/HCT 56)

The answer to the question, as we have seen, lies with the notion of categorial intuition. Interestingly, Heidegger then goes on to sharpen the distinction between sensuous and categorial intuition with reference to the distinction between *simple* and *multi-level acts* discussed in Paragraph 46 of the sixth *Logical Investigation*.

This distinction can be expressed visually thus:

(real) – simple acts = low-level = founding = sensuous intuition
(ideal) – complex acts = multi-level = founded = categorial intuition

In Husserlian terms, simple acts are founding or low-level acts, where an object is simply given to perception. Multi-level acts are those founded or complex acts, where an object is articulated through an assertion or categorial intuition. Now, although *de jure* the distinction between simple and multi-level acts can be made, it is always the case *de facto* that the founding moment of simple sensuous intuition is articulated through an assertion that employs multi-level categorial forms—the sensuous is always already shot through with the categorial. Hence, the relation between simple and multi-level acts or sensuous and categorial intuition is characterized by *interdependence*, where founded categorial acts—the *being*-brown of the chair—are dependent upon founding perceptual acts—the *brownness* of the chair—but where the founding only becomes accessible, one might even say meaningfully visible, for the first time in the founded articulation that takes place in the assertion.

Heidegger writes: "...the founded acts *disclose anew* [*neu erschliessen*] the simply given objects, such that these objects come to explicit apprehension precisely in what they are" (PGZ 84/HCT 62). Sensuous intuition is shot through with categorial acts that meaningfully disclose the sensuous *as* sensuous—seeing is a saying.

Heidegger then goes on to distinguish two groups of such categorial, founded acts: *acts of synthesis* and *acts of ideation*. In acts of synthesis—and Heidegger is here engaging in a discreet retrieval of the functions of *synthesis* and *diairesis* in Aristotle's *Peri hermeneia* (*On Interpretation*), a decisive text for his conception of language and hermeneutics as can be seen from Paragraph 33 of *Sein und Zeit* (SZ 159)—the founding, sensuous or real intuition and the founded, categorial or ideal intuition are synthesized in the assertion. In acts of ideation, the founding, real moment is left behind and the focus is the purely ideal categorial form. For example, the assertion "the chair is brown" enacts a synthesis of the ideal and the real, or the sensuous and the categorial, such that the real is expressed or articulated through the ideal and becomes meaningfully visible for the first time. The founded, ideal or categorial forms of the assertion take up and express the founding, sensuous or real elements and synthesize them into a total meaning-situation that discloses those objects anew.

In acts of ideation, if only *de jure*, the founding real act is *diairetically* separated from the founded ideal act, what Husserl calls at the end of the sixth *Logical Investigation*, "the intuition of the universal" (*der Anschauung des Allgemeinen*), what he calls in his later work eidetic intuition.[30] As Heidegger makes clear, what is being retrieved here is the Platonic notion of *eidos* understood as "the outward appearance of something" (*das Aussehen von Etwas*), a translation that appears in Heidegger's critique of epistemology in Paragraph 13 of *Sein und Zeit*, where *eidos* is a pure *Aussehen* that arises as a deficient mode of Dasein's *Benommenheit*, its benumbed or fascinated being-at-home-in-the-world (PGZ 90/HCT 66, SZ 61). What seems to interest Heidegger here is the way in which, at the end of his discussion of categorial intuition, Husserl is led to claim an intuition of the universal as something beheld in the assertion's act of meaning. Ideation is the act that gives the *eidos*, what Cobb-Stevens calls, with Aristotle, "the species-look" of a thing.[31] Although the universal must be based upon a founding, perceptual moment, in the act of ideation it is the abstraction of the founded, categorial moment that I focus on: I do not see the brownness of this particular chair as much as brownness in general.

In this way, Heidegger is led into the thorny philosophical dispute about the status of universals in the medieval debate between realism and nominalism, where the realist argues that universals have substantial, real existence, and the nominalist claims that only sensuous particulars are real and therefore universals have no existence independent of thought—they are mere names. With a gesture that is familiar to us from the above discussion of Kant, Heidegger agrees with the nominalist critique of universals insofar as they are not the real, substantial constituents of objects, but he does not want to be led from this critique into the denial of their objective existence. Rather, the phenomenological (re)discovery of categorial intuition permits philosophy to focus once again on the nature of the universal, the *eidos* or the a priori as that which is seen in an ideative act. That is, in a way that will become clearer presently, philosophy can raise anew the question of the meaning of being as its fundamental matter.

III

In his analytically sharp but somewhat impatient book on Husserl, David Bell makes the following remark about the notion of categorial intuition: "I have been unable to find evidence in Husserl's text that is clear enough to enable me to say with any confidence just what his theory on these matters is."[32] I think Bell goes too far, too fast, but we saw above that the skeptical questions that can be raised against the notion of categorial intuition necessitate some sort of independent reconstruction of Husserl's argument. I would like to present this in eight steps.

1   For Husserl, like Brentano, each psychic phenomenon is either a presentation or is founded on a presentation. That is, sensuous intuition or perception is the founding moment of all knowledge and meaning, and the latter would be unintelligible without perceptual experience. As Heidegger says, restating

Husserl from the sixth *Logical Investigation*, "A thought without a founding sensuousness is absurd" (PGZ 94/HCT 69).[33] In Kantian terms, concepts without intuitions are blind.

2 But, in order for perceptual experience, simple seeing, to be rendered meaningful or intelligible, it has to be articulated through linguistic expressions: for Husserl, in assertions or propositions; for Heidegger, in pre-propositional talk or *Rede*. Thus, language is the condition of possibility for the meaningful visibility of perceptual items (there can, of course, be meaningless visibility).

3 The position described in (2) is just *how it is* for Husserl, insofar as we are meaning-bestowing beings in a world that is, for us, meaning-full. We give expression to experience in language and literally *make* sense. This is something we just do; that is, experience begins—as it does for Kant—from the fact of the synthesis of concepts and percepts.

4 Insofar as we do this, perceptual experience or sensuous intuition is shot through with or completely pervaded by linguistic expressions and categorial acts. Furthermore, this pervasion of perception by language—of seeing by saying—is the condition of possibility for relatively unambiguous communication, for the sharing of experience among speakers of a common language, which recalls Heidegger's point about the *kategoriai* as a public matter. As I noted above, in infant language acquisition, the ambiguity of the word "ball" is reduced by the insertion of categorial forms. Or again, the assertion "this is a brown chair" only becomes meaningfully articulated and understandable through the mediation of the categories "this", "is", "a", "brown"; that is, of location, substance, quality and quantity; and through the creation of the abstract noun, "chair".

5 So, what I am saying here is that the process whereby perceptual experience is rendered intelligible or meaningful is one of abstraction, generalization or ideation. For example, if I see three chairs in front of me C1, C2 and C3, all of which are of the same color, let's call it x, then x is the abstract or ideal species of color that all these perceptual items have in common. Thus, one can say, and this is crucial to the argument, the species or universal x is something *seen*. My awareness that all those chairs possess the color x is a perceptual awareness; that is, I can see the color that C1, C2 and C3 have in common, but not in the same way as I can see the particular chairs. If, suddenly, someone showed me that the chairs could be seen as something else, as a different shade of color, as y and not x, then the change in the species of color would be a change in perceptual awareness, e.g. I see the chair as brown and then as gray-green. Think of the celebrated Wittgensteinian example of the duck-rabbit, where I see something as a duck and then as a rabbit. For Wittgenstein, the change of aspect or modification in perception that takes place in seeing something first as a duck and then as a rabbit, is not a change in my opinion, judgment or inner state, but a change at the level of perceptual awareness. As Wittgenstein says, "the concept of seeing is modified here".[34] I think that this is what Heidegger has in mind when he

says, "the characterization of ideation as a categorial intuition has made it clear that something like the highlighting of the ideas [*die Abhebung von Ideen*] occurs both in the field of the ideal, hence of the categories, and in the field of the real" (PGZ 101/HCT 74).

6   Thus, perceptual experience is rendered intelligible through a process of ideation, generalization, or abstraction that occurs with linguistic expression. In language we move from the particular to the universal, from the brown of this particular chair to brownness as such; or, crucially, for Heidegger, the *being*-brown of the chair. The important thing here is that this move from the particular to the universal takes place at the level of intuition, it is something *seen* in the object and not a subjective construction. Categorial forms are not the subjective constructs of acts but the objects that manifest themselves in those acts. Thus, although there is no return to a pre-modern *intuitus originarius* in Husserl, one can say that there is an intuition of the categorial as a founded aspect of the total meaning-situation.[35] The move from sensuous to categorial intuition is a little like undergoing aspect change in Wittgenstein, where I can shift my attention—or have my attention shifted by another—from the sensuous to the categorial moment in the total meaning-situation. I can shift from this particular brown (being-*brown*) as a particular sensuous item, to brownness as an abstract sensory idea (*being*-brown), and ultimately, in an act of ideation, to the being of brownness as a pure categorial form without regard to its sensuousness (*being*).

7   One final proviso: this entire analysis, as my discussion of acts of ideation and synthesis pointed out, is presupposed upon a *de facto/de jure* distinction: that is, *de facto* sensuous intuition is articulated and rendered meaningful through acts of synthesis where it is combined with categorial, linguistic forms. The two elements in the act of synthesis always come together, they are inextricably linked and immersed in each other. The founded, categorial moment is the condition of possibility for the intelligibility of the sensuous, but the sensuous is the founding condition of possibility for there being something for the categorial to render intelligible. This is how it is *de facto*; however, *de jure*, one can distinguish between the sensuous and the categorial and rest one's focus on the real or ideal moment in the total meaning-situation. It is in this sense, and only in this sense, that one can plausibly speak of an intuition of the universal, the *eidos* or species-look, or the a priori. *De facto*, there is no such thing as an intuition of being.

8   Looking ahead to both the later parts of these lectures and the opening distinctions of *Sein und Zeit*, it would then be a question of recasting this entire discussion of the objectivity of categorial forms into the language of the *existentials*, namely those a priori general features of beings like us, who are defined by a "who" and not by a "what". That is, it would be a question of deconstructing the massive privileging of *Vorhandenheit* that is everywhere at work in the discussion of categorial intuition and mobilizing the formal-methodological concept of phenomenology in a concrete-hermeneutics of

everyday being-in-the-world. This, at least, is the huge ambition of *Sein und Zeit*. As Heidegger remarks in that volume, "Even the phenomenological *Wesenschau* is grounded in existential understanding" (SZ 147).

The phenomenological a priori

The discussion of the a priori occupies considerably less space than that given to the two previous discoveries of phenomenology—a mere three pages—and essentially functions as a coda to the positive reading of Husserl and a transition to the definition of the formal-methodological concept of phenomenology. The structure of Heidegger's argumentation follows what is by now, I hope, a familiar pattern.

It is claimed that since Kant (which means, of course, in Heidegger's periodization of modernity, since Descartes [PGZ 100/HCT 73]), the a priori has been exclusively identified with knowledge, and thus defined in entirely epistemological terms. For Kant, the a priori is characterized by universality and necessity and is a term applied to all judgments whose validity is independent of any empirical, a posteriori data, which are by definition particular and contingent. The Kantian categories, as the pure concepts of the understanding and the transcendental conditions of possibility for uniting the manifold of intuition, are a priori. This leads Heidegger to a second claim, namely that just as there was a subjectivization of the categorial in Kant, so too there is a subjectivization of the a priori. What this means is that the a priori is, for Kant, a feature of judgment, obviously and particularly synthetic a priori judgments, and that judgment is a feature of the subjective sphere.

Of course, Heidegger is treading on philosophical ice that is so thin one can almost hear it cracking under his feet. This is particularly the case when one thinks of what he will claim about the schematism in *Kant and the Problem of Metaphysics* in 1929. In the latter, Heidegger argues that Kant had an insight into the temporal nature of being but shrank back from what he saw there.

Against this coupling of the a priori and the subject, Heidegger wants to claim that the Husserlian theses on intentionality and categorial intuition, specifically the previously discussed concept of ideative acts where categorial forms are given trans-subjectively in intentional experience, show that "...the a priori is not limited to subjectivity, indeed that in the first instance it has primarily nothing at all to do with subjectivity" (PGZ 101/HCT 74). This interpretation of Husserl would appear to be justified in the light of the above-cited remarks from the end of the sixth *Logical Investigation*, where it is claimed that Kant "...drops from the outset into the channel of a metaphysical epistemology". Husserl goes on to claim that Kant "...never made clear to himself the peculiar character of pure ideation, the adequate survey of conceptual essences, and of the laws of universal validity rooted in those essences. *He accordingly lacked the phenomenologically correct concept of the a priori*.".[36]

As Husserl points out in *The Idea of Phenomenology*—a text from 1907, but not published until 1950—a priori cognition directs itself towards general essences, it is a phenomenological seeing of the a priori in an act of absolute self-givenness.[37]

For Heidegger, Husserlian phenomenology achieves a *de-subjectivization of the a priori*, a retrieval of the original sense of the a priori as the intuiting of the universal in the sense outlined above. In the *Basic Problems of Phenomenology*, Heidegger refers to Plato as "the discoverer of the a priori", and in this sense Husserl, as we have already seen, unwittingly recalls the Platonic notion of the *eidos*. In an important early footnote to *Sein und Zeit*, Heidegger stipulates that "'a priorism' is the method of every scientific philosophy that understands itself" (*die sich selbst versteht*) (SZ 50). In *Sein und Zeit*, however, the a priori horizon for the analysis of *Dasein* is being-in-the-world as it is disclosed in the mode of average everydayness, and what Heidegger seeks to do with his notion of *existentials* is to describe the de-subjectivized a priori structures of Dasein's intelligibility of world and self.[38]

For Heidegger, the phenomenological a priori is not a feature of the subjective sphere, but "*ein **Titel des Seins***" ("a *title for Being*", PGZ 101/HCT 74). To understand this claim, it is necessary to see the "temporal" nature of the a priori, its etymological reference to the *prius* and the *proteron*, the prior and the "before". For Heidegger, the a priori is *das Frühere*, it implies a time sequence in which the prior is "the earlier" and more original because it is earlier (PGZ 99/HCT 73).[39] The a priori is that which comes before.

Although Heidegger just makes a passing reference in the *Prolegomena* to the temporal sense of the a priori, this can be glossed with reference to the *Basic Problems of Phenomenology*, where inquiry into the meaning of being is characterized as a priori in the sense that being has ontological priority over particular beings, even if access to the *Seinsfrage* must first be secured by starting with some being, some posterior, namely Dasein. To expand the focus for a moment, if Heidegger's leading claim in *Sein und Zeit* is that the meaning of the being of Dasein is finite temporality (*Zeitlichkeit*), then the fundamental ontological task still outstanding (as is made clear at the end of Paragraph 83) is the elaboration of the temporal sense of being itself, what Heidegger names with the expression *die Temporalität des Seins*, the temporality of being that he tries to capture in his later work with the notion of *das Ereignis*. The point here is that for Heidegger the temporal sense of being as such is the "earliest" and most original of all matters, and therefore the authentic a priori of phenomenology.[40]

To summarize, Heidegger is making a threefold claim with regard to the phenomenological a priori:

- first, that it is universal and indifferent to subjectivity;
- second, that it is not a feature of a subjective judgment, but is an aspect of intuition, namely Husserl's notion of categorial intuition;
- third, that the authentic a priori is being, whose meaning is to be understood in temporal terms.

Phenomenology as renewal

I would now like to pull together the strands of the first moment of the double gesture that governs Heidegger's reading of Husserl by returning to what I called above the "nesting effect" of the three discoveries of phenomenology and see how they are synthesized into a new definition of the task of phenomenology. The positive reflection on Husserl gives an almost archeological account of phenomenology, beginning with intentionality and then digging into the a priori structure of intentionality with the account of categorial intuition, before hitting bottom with the phenomenological a priori, which allows the *Seinsfrage* to emerge as the authentic matter of phenomenology.

It should be emphasized once again that these three discoveries find their methodological condition of possibility in the first discovery. Intentionality is the essential structure of the subject; which means, for Heidegger, that intentionality de-subjectivizes the human and shows how mental experience has the character of intentional comportment towards things. The being of being human is intentional; it is a *Sich-richten-auf* that will allow Heidegger to define the basic trait of Dasein as transcendence towards the world. Categorial intuition is the research into the a priori structure of intentionality; in other words, it asks the question of the being of the intentional and defines the field of phenomenological research as "*intentionality in its a priori*" (PGZ 106/HCT 78). As we have seen, the Husserlian concept of categorial intuition allows Heidegger to define the phenomenological a priori in terms of the trans-subjective self-givenness of being in an act of phenomenological seeing, a seeing which is also a saying.

Now, if being is given in an act of seeing, or is an aspect of the manifold of intuition, what Heidegger calls "a simple originary apprehension" ("*...schlichten originären Erfassens*", PGZ 107/HCT 78), then this means that the correct mode of treatment for this apprehension is a description (*Beschreibung*)—the phenomenologist describes what s/he sees. Obviously, Husserl's notion of description and of phenomenology as a descriptive psychology is transformed into Heidegger's notions of interpretation and hermeneutic in *Sein und Zeit*, although the notion of hermeneutic is at work earlier in Heidegger, for example in his 1923 lecture course *Ontology—The Hermeneutics of Facticity*.[41] Although the move from description to interpretation is obviously significant, the gulf between descriptive and hermeneutic phenomenology is not as great as certain commentators, such as Hubert Dreyfus, would like to imagine.[42] Heidegger goes on to claim that such phenomenological description takes place in an *analytic*, a word employed in a seemingly Kantian sense to specify the nature of a priori research into the categories. This leads Heidegger to the rather baroque definition of phenomenology as the "*analytic description of intentionality in its a priori*" (PGZ 108/HCT 79).

The three discoveries of Husserlian phenomenology culminate in a renewed sense of the task of phenomenology and of philosophy itself. It is my contention that, by following the first part of the double gesture of the *Prolegomena*, one can understand how Heidegger can at least consistently and coherently claim, as he

31

does in the introduction to *Sein und Zeit* that the proper matter of phenomenology, its *Gefragte* (that which is asked about), is the being of beings (SZ 37). This entails, of course, that phenomenology becomes ontology, or vice versa, that "ontology is only possible as phenomenology" (SZ 35). This is also to say that phenomenology is the *science* of being in the Aristotelian sense, i.e. metaphysics. Furthermore, by insisting that phenomenology's basic theme is intentionality, Heidegger establishes the coherence of the claim that access to the *Seinsfrage* must first pass through an analysis of that being whose fundamental trait is intentional comportment; that is to say, through a fundamental ontology of the *Befragte* (that which is interrogated), i.e. Dasein. Only such a Dasein-analytic will permit access to the *Erfragte* (that which is to be found out by the asking) of Heidegger's work, the meaning of being. With this in mind, one can see the sense of Heidegger's apparently grandiloquent claim at the end of the introduction to *Sein und Zeit*, that '...philosophy is universal phenomenological ontology that takes its departure from a hermeneutic of Dasein and always returns to that point of departure" (SZ 38).

I would like to make some skeptical remarks before continuing.

1   Heidegger's subtle but significant modification of the rallying cry of phenomenology should be noted. Husserl's maxim *"Zu den Sachen selbst"* (SZ 27, PGZ 75/HCT 103) becomes, in the later work *"Zur Sache selbst"*.[43] This seemingly minimal alteration of the plural 'matters' to the singular 'matter' *matters* because it presupposes the entire future orientation of the Heideggerian enterprise, and its difference from that of Husserl. Whereas the former is concerned with elaborating the single *Seinsfrage*, the latter begins by trying to account for the plurality of matters experienced by consciousness, and consciousness for Husserl is an extremely broad notion, a sort of bucket term for the location of mental experiences. Interestingly, in this connection, Klaus Held attempts to reconcile the difference between Husserl and Heidegger by claiming that they were both trying to think *eine Sache*, namely the world as the unifying horizon for the plurality of appearances.[44] However, if this claim is plausible—and I think it is extremely tempting—then the question of being as something distinct from the question of world is not tenable. Or, better, that the question of being is simply a way of raising the question of world with a phenomenological radicality that questions the way in which the world has been conceived and covered over in the modern philosophical tradition, Husserl included. One might conclude that all talk of being independently from the being of the world is simply redundant, or is what I like to call a red herring.

2   On the question of whether the *Seinsfrage* must begin from a Dasein-analytic, and hence from fundamental ontology, some account must be taken of Heidegger's excruciatingly honest doubts about the enterprise of fundamental ontology in the final paragraph of *Sein und Zeit*, where the Dasein-analytic is characterized as *"nur ein Weg"* to the *Seinsfrage* (SZ 436)

although given Heidegger's extreme valorization of the word *Weg* as that which best describes the itinerant movement of thinking, it is difficult to know what the qualification "only" would mean here). Also, it would have to be noted how, after the publication of *Sein und Zeit*, beginning in the *Basic Problems of Phenomenology* and predicting the entire future trajectory of the so-called *Kehre*, Heidegger's thinking is orientated around the question of being as it is articulated and received historically in the tradition and is not pursued through the quasi-anthropology of an existential analytic. A polemical question: can one retain the philosophical paradigm shift involved in the project of fundamental ontology without necessarily signing up to Heidegger's account of the history of being? Perhaps. This raises a huge issue around the relation between the phenomenologically constructive and destructive elements in Heidegger's work. Let me elaborate.

3   My claim that the beginning of Heidegger's thought is phenomenological opens itself to the objection that such an approach plays down the *destructive* or *deconstructive* side to Heidegger's project.[45] That is, as Heidegger puts it at the end of Paragraph 6 of *Sein und Zeit*, "The question of being does not achieve its true concreteness until we have carried through the destruction of the history of ontology" (SZ 26). On this understanding, the beginning of Heidegger's philosophy is found in his repetition or retrieval (*Wiederholung*) of the question of being as it was first articulated by the Greeks in the ontology of Plato and Aristotle. This is why the text of *Sein und Zeit* begins on the untitled first page with a quotation from Plato's *Sophist*. This is an important objection, but let me clarify what I am trying to do in these lectures. I am seeking to analyze and, if possible, justify, the formal-methodological concept of phenomenology at work in Heidegger. Now, such an approach undoubtedly needs to be de-formalized, to use Heidegger's word (*entformalisiert*, SZ 35), through both the specific phenomenological analyses of *Sein und Zeit*, and the destruction of the ontological tradition, if the concreteness of which Heidegger speaks above is to be achieved. Therefore the phenomenological approach I am recommending has to be *complemented* by a destructive or deconstructive approach in terms of Heidegger's engagement with the philosophical tradition. For example, Heidegger's strategy with regard to the three discoveries of phenomenology in the *Prolegomena* is to locate the point where each of these concepts crosses the path of the ancient ontology of Plato and Aristotle. Ultimately, the trans-subjective givenness of being expressed by the doctrine of categorial intuition allows Heidegger to reactivate the Greek determination of being as presencing (*Anwesenheit*, SZ 26), and hence to reawaken the link between being and time. The deformalization of the phenomenological approach is achieved, for Heidegger, by way of a repetition of the Greek beginning of philosophy, what he calls in the *Prolegomena* the "assumption of the tradition as a genuine repetition" (PGZ 187/HCT 138). However, my ambition is simply to analyze the formal-methodological

tools that permit this deformalizing assumption of tradition. By itself—I would insist—tradition can and should assume no authority in philosophical matters.

Phenomenology as tautology

The positive reflection on Husserl in the *Prolegomena* concludes with a clarification of the name "phenomenology", a discussion that is copied, expanded and reworked as the crucial methodological Paragraph 7 of *Sein und Zeit*. The definitions of "phenomenon" and "logos" are extended in *Sein und Zeit*, with references to Kant being added to the former (SZ 31), and important remarks on truth and judgment being added to the latter (SZ 32–34). However, in the light of my line of argument in these lectures, what is most curious and, dare one say, intellectually misleading, about Paragraph 7 is that, apart from the previously discussed reference and footnote dedicated to Husserl (SZ 38, which could be glossed as an obsequious but peripheral genuflection to a former teacher, which is certainly how I first read it as an undergraduate), all substantive reference to Husserl has been excised from the definition of phenomenology in *Sein und Zeit*. Now, the only difference between the claim in the *Prolegomena* that the matter of phenomenology is intentionality in its a priori and the claim in *Sein und Zeit* that phenomenology is a science of the being of beings, is that the former is attained through an explicit interpretation of Husserl, while the latter elides—but implicitly presupposes—such an interpretation. This elision can already be seen to be prepared in the lecture course that directly precedes the *Prolegomena*, in the 1924–25 lectures on the *Sophist*, where the essentials of the definition of phenomenology in *Sein und Zeit* are already in place. Heidegger defines "phenomenon" as *sich zeigen* and "logos" as *Ansprechen*, speaking to, which also has the sense of an appeal or an address, even a demand, an *Anspruch*, and manifests the same ambivalence towards Husserl. It is claimed that the *Logical Investigations* are the first breakthrough in phenomenology, but Heidegger already gives a coded critique of the Husserlian privileging of consciousness and says that phenomenology cannot be learnt by reading the literature of the phenomenological movement, but by "*die Arbeit des Durchsprechens der Sachen*", "the work of speaking through things".[46]

It is, of course, conjectural to what extent one's perspective on *Sein und Zeit* (and indeed Heidegger's work as a whole) would be transformed if one rejoined it to its buried phenomenological preface. But it can at the least be emphasized that the interpretative stakes here are rather high and, arguably, the interpretation of Heidegger turns on how one reads Paragraph 7, and what prominence one gives to Heidegger's Discourse on the Method, as it were. To emphasize my point, let me take up three examples from the literature on Heidegger by a German, an American and a Frenchman.

In his minute and distressingly exact commentary on the introduction to *Sein und Zeit*, von Herrmann devotes no less than 115 pages to Paragraph 7.[47] He argues that Heidegger's allegiance to Husserl is entirely methodological and not at all thematic; that is, Heidegger and Husserl have widely different conceptions of the theme of phenomenology, *Bewußtsein* or consciousness for the former and *Dasein* for the latter. As I have tried to show, such a reading is inadequate and simply obviates the way in which Heidegger takes up and transforms the thematic, substantive concerns of Husserlian phenomenology, such as intentionality.

However, what von Herrmann says in 115 pages, Hubert Dreyfus says in three, which is all he devotes to Paragraph 7 in his influential commentary on the First Division of *Sein und Zeit*.[48] For Dreyfus's broadly pragmatist interpretation of Heidegger to be effective, it is essential that he separate out Husserl's allegedly "epistemological, foundationalist inquiry" from Heidegger's supposedly "ontological, pragmatist anti-foundationalism". This entails that Heidegger's conception of phenomenology means "exactly the opposite of Husserl".[49] Dreyfus reinforces this opposition through a sharply drawn distinction between Husserlian descriptive and Heideggerian hermeneutic phenomenology. Needless to say, as I have tried to show, it is the purpose of these lectures to challenge such a strict division between Husserl and Heidegger.

A third interpretation of Paragraph 7, one that I find much more plausible, is given in Jean-François Courtine's *Heidegger et la phénoménologie*.[50] Courtine describes Paragraph 7 as *le coup d'envoi* or the kick-off to *Sein und Zeit*, and claims that all the preceding paragraphs culminate in the definition of the *Vor-Begriff* of phenomenology. Heidegger's definition of phenomenology in terms of possibility determines, for Courtine, the entire future orientation of Heidegger's thinking insofar as the treatment of the *Seinsfrage* is only possible phenomenologically. This entails, against both von Herrmann and Dreyfus, that Paragraph 7 is not a simple methodological excursus, and to treat it as such is simply to engage in a strategic reception of Heidegger. The guiding assumption of these lectures is to try and *understand* the entirety of the methodological ambition of Heidegger's work and to do so *systematically*. It is only on the basis of a systematic reconstruction that Heidegger's work can be convincingly criticized.

If we now let the light gleaned from the reading of Husserl in the *Prolegomena* fall across the face of Paragraph 7, several hitherto obscured features are brought more sharply into relief. First, the discussion of categorial intuition sheds light on Heidegger's conception of the phenomenon. Heidegger, it will be recalled, traces "phenomenon" back to the Greek verb *phainesthai*, which he then retranslates as *sich zeigen*. "Phenomenon" is defined as that which shows itself in itself (*sich-an-ihm-selbst-zeigende*, SZ 28). In a gesture that one can also find in the *Sophist* lectures, Heidegger then claims that this conception of phenomenon is distinct from, and transcendentally grounds, the "vulgar" concepts of phenomenon such as *Schein* (shine or semblance) and *Erscheinung* (appearance). What Heidegger means by "phenomenon" is not therefore assimilable to Husserlian sensuous intuition, Kantian empirical intuition or natural perception. Rather, the phenomenon is categorial and the categorial forms are phenomenal; that is, it is being itself that is phenomenal insofar as it shows itself in itself non-sensuously and non-empirically in the self-givenness of categorial intuition. This explains how Heidegger is able to make the apparently counter-intuitive claim that what he means by "phenomenon" can be equated with what Kant means by space and time as the a priori forms of intuition (SZ 31), but note that Heidegger takes his Kantian example from the transcendental aesthetic where it is a question of the a priori forms of intuition, rather than the deduction of the categories in the transcendental analytic because the latter is already, for him, overtaken by a questionable conception of the subject.

Second, the above discussion of the nature of the assertion in the *Prolegomena* prepares the way for the concept of *logos*. The latter is translated as *Rede* or talk, whereas in the 1924 lecture *Der Begriff der Zeit*, Heidegger renders it as *das Sprechen*, or speech. We should also note the way Heidegger deals with this issue in the *Sophist* lectures, where he claims that the Aristotelian notion of *logos*, which is there rendered as *Ansprechen* (speaking to or address) is rediscovered by Husserl in the intentionality thesis, where "*jedes Ansprechen ist Ansprechen von Etwas*", "every address is an address about something".[51] Heidegger defines the function of talk as making manifest what the talk is about, a function that he describes with the verb *apophainesthai*, which is translated as *lassen sehen*, to let see. Talk therefore lets us see what it is that we are talking about, it has a necessarily disclosive function, as Heidegger puts it in Paragraph 34 of *Sein und Zeit*: "*der Mensch zeigt sich als Seiendes, das redet*" ("the human being shows itself as the being that talks", SZ 165). But what does talk let us see? Recalling the discussion of categorial intuition, the saying of the assertion allows us to see the matter of phenomenology, the categorial form of being is disclosed linguistically as an aspect of intuition.

Third, Heidegger unifies the meaning of "phenomenon" and "logos" into the tautological definition of the *Vor-Begriff* of phenomenology as *apophainesthai ta phainomena*, translated as "to let see (*sehen lassen*), that which shows itself (*was sich zeigt*)" (SZ 34). What shows itself in phenomenology, what it enables us to see, Heidegger claims, is being itself (SZ 35), and the way in which it is disclosed is in language, that is to say, through the non-sensuous categorial seeing of the being that shows itself in the saying of the copula. The fact that the definition of phenomenology is a tautology is not simply accidental, and Courtine has persuasively argued that there is a "transmutation" of phenomenology into tautology in Heidegger's work.[52] Heidegger untranslatably renders the Greek *apophainesthai ta phainomena* as "*was sich zeigt, so wie es sich von ihm selbst her zeigt, von ihm selbst her sehen lassen*" ("to let that which shows itself be seen from itself in the very way in which it shows itself from itself", SZ 34).

The philosophical point of such tautologous formulae is that *if* the truth of phenomenology—which for Heidegger is ultimately the verbal sense of being as presencing (*Anwesenheit*)—cannot be expressed propositionally, in sentences using the copula, then it can perhaps be best expressed tautologically where substantives become verbalized, and where *that which is* is not conceived as a substance but rather as a temporal process. This is something that is not just present in the tautologous definition of phenomenology. It can also be observed in Heidegger's predilection for tautological formulae, for example: *die Welt weltet, das Ereignis ereignet,* and *die Sprache spricht* (the world worlds, the appropriative event appropriates, and language speaks). It can also be seen in the entire trajectory of the *es gibt* in Heidegger, whether it is employed as a way of avoiding the use of the copula in propositions—*es gibt Sein, es gibt Zeit*—or whether it is traced into the deceptive simplicity of Georg Trakl's lyric poems or Rimbaud's repeated use of the *il y a* in *Les illuminations.*[53] It can also be seen in Heidegger's attempts at poetry in *Aus der Erfahrung des Denkens*:

> Wälder lagern
> Bäche sturzen
> Felsen dauern
> Regen rinnt.
>
> Fluren warten
> Brunnen quellen
> Winde wohnen
> Segen sinnt.[54]

What tautologous formulations articulate—precisely in saying nothing as the early Wittgenstein would put it—is the verbality of a pre-given, but ultimately opaque and enigmatic, facticity. We will come back to this thought below when we turn more closely to *Sein und Zeit.*

The possibility of falling

As Heidegger insists—and this insight will inform his entire future account of the care-structure, the structure of *aletheia* and the withdrawing-donating movement of the history of being—the phenomenon with which phenomenology deals runs the risk of becoming covered up (*verdeckt*) and even the most concrete work of phenomenology runs the continual risk of entering crisis by letting its matter become hardened.

Thinking of what Heidegger will see as the fate of the question of being in Husserl's *Ideas I* (to which I will turn presently), the real difficulty of phenomenological research is making it critical in a positive sense (SZ 36). Borrowing a metaphor from Husserl's *Crisis of the European Sciences*, phenomenology must be a permanent *reactivation* of critique, where the phenomena need to be perpetually wrested from the temptation of what Husserl calls "sedimentation", and what Heidegger more dramatically calls "falling" or "decadence" (*Verfall*).[55] Phenomenology as a *passion for absolute philosophical radicality* (and I shall have reason to question this passion below) must confront the temptation of entering crisis by engaging in a process of permanent critical renewal. I think this is why Heidegger writes: "Our comments on the preliminary concept of phenomenology have shown that what is essential in it does not lie in its *actuality* as a philosophical 'movement' [or 'trend', *Richtung*]. Higher than actuality stands *possibility*. We can understand phenomenology only by seizing upon it as a possibility." (SZ 38).

Phenomenology, as Heidegger tirelessly reminds us, is not the name of a philosophical movement or a past trend in the history of ideas, it is rather *the permanent renewal of its own possibility as possibility*. If phenomenology is essentially defined in terms of possibility, then by definition it is incomplete and future-directed. That is, phenomenology is, by definition, something for which there can only be a preliminary concept, and not the full concept of phenomenology, momentarily anticipated in Paragraph 69(b) of *Sein und Zeit* (SZ 357). As Heidegger writes at the end of his critical reflection on Husserl, with characteristic but not accidental hyperbole:

> The greatness of the discovery of phenomenology lies not in factually obtained results, which can be evaluated and criticized and in these days have certainly evoked a veritable transformation in questioning and working, but rather in this: it is the *discovery of the very possibility of doing research in philosophy*. But a possibility is rightly understood in its own most proper sense only when it continues to be taken as a possibility and preserved as a possibility.
>
> (PGZ 184/HCT 135–36)

Phenomenology, for Heidegger, is a methodological concept that specifies not the what (*das Was*) but the how (*das Wie*) of philosophical research.

Phenomenology is the very possibility of philosophy as a tendency to keep open to the matters themselves, to reactivate the origins of a practice or a state of affairs against the hardening process of sedimentation that substitutes traditionality for radicality, and crisis for critique. This emphasis on the task of philosophy as possibility is something that one might also hear (despite their vast differences of tone, temper and tradition) in Wittgenstein's remark from the *Philosophical Investigations*: "Es ist uns, als müßten wir die Erscheinungen **durchschauen**: unsere Untersuchung aber richtet sich nicht auf die **Erscheinungen**, sondern, wie Man sagen könnte, auf die '**Möglichkeiten**' der Erscheinungen".[56]

Phenomenology is not, therefore, a region of philosophy, a branch of the tree of metaphysics, but is rather the very possibility of philosophy, of philosophy as the possibility of possibility. Defined in these terms, phenomenology has the ambition of freeing us from the bonds of tradition—specifically the Cartesian tradition that accords primacy to the subject and privileges the ontology of *Vorhandenheit*—and permits what Heidegger calls a retrieval (*Wiederholung*) of the radical beginning of philosophy, that is to say, a renewal of the questioning of Plato and Aristotle (PGZ 184/HCT 136). And this is where *Sein und Zeit* begins with its untitled first page and its quotation from Plato's Sophist: "For manifestly, you have long been aware of what you mean when you use the expression 'being'. We, however, who used to think we understood it, have now become perplexed".[57]

The passage to philosophy begins as a movement into aporia or perplexity. For Heidegger, it is this movement into perplexity that is reactivated by phenomenology and which must be kept open as a possibility by relentlessly restating the *Seinsfrage*.

So, to return to the questions with which I framed these lectures, it should now hopefully be clear *where* one should begin with Heidegger and *what* is the beginning of his philosophical project. What is not yet clear is *why* one should begin philosophizing with Heidegger rather than elsewhere. I would like to approach this question by taking a slightly different tack.

Transforming the natural attitude—from personalistic psychology to Dasein analytic

Let me turn to the other side of the double gesture in Heidegger's reading of Husserl, the tenor of which also remains unchanged in his later work and upon which I would like to concentrate for the remainder of these lectures. The general claim here is that if Husserl's notion of categorial intuition is the *Boden* upon which the question of the meaning of being can be raised as a substantive philosophical issue, then after the publication of the *Logical Investigations* in 1900, Husserl failed to pursue the *Seinsfrage* with sufficient radicality. The publication of the first volume of Husserl's *Ideas* in 1913 constitutes, for Heidegger, a philosophical *decision* to sacrifice radicality for traditionality.

We should note this pairing of terms in Heidegger's work of this period, where what is continually valorized in philosophy (and in much else, it would appear) is an absolute *radicality* whose antonym is *tradition*. Heidegger's work—and this is hardly a neutral matter, particularly when one thinks of the completely overdetermined philosophical and political thematics governing the language of the decision (*Entscheidung*) in the Germany of the 1920s—is motivated by a passion for absolute philosophical radicality. As is common in Heidegger, tradition is always understood in terms of the Cartesian legacy of the modern determination of being as subjectivity. Husserl's traditionality is therefore synonymous with his alleged Cartesianism, where the phenomenological field in *Ideas I* is constituted as a realm of pure consciousness, and where the latter is determined as absolute being, whose investigation is the subject matter of a rigorous science: transcendental phenomenology.

Heidegger takes a rather malicious delight in referring extensively to Paragraphs 46–50 of *Ideas I*, where consciousness is determined as indubitable, pure, absolute and immanent being in opposition to the dubitability, relativity and contingency of reality, and where Husserl famously claims that consciousness would be *modified* (indeed!) by the nullification of the world, but not affected in its own existence.[58] But the core of Heidegger's critique of the later Husserl is that in determining the phenomenological field as that of pure consciousness, he fails to pose the question of the being of consciousness, or what Heidegger calls the being of the intentional (*das Sein des Intentionalen*), and consequently loses sight of the *Seinsfrage*. In other words, in determining pure consciousness as absolute being, Husserl takes over a conception of consciousness from the tradition without interrogating its meaning.

If this claim is justified—and I am not saying that it is, as Heidegger's reading of Husserl's work after the *Logical Investigations* is extremely myopic—then this explains why Heidegger goes on to claim that the Husserlian notion of consciousness is *unphenomenological* insofar as it is not drawn from the matters themselves, i.e. from the lived experiences of a concrete human being, from *was sich selbst zeigt*, but is inherited from the tradition, specifically the Cartesian tradition (PGZ 147/HCT 107). Thus, for Heidegger, Husserlian phenomenology becomes unphenomenological, it sacrifices radicality for traditionality.

Now, if this is the fate of Husserlian phenomenology, then the question becomes: how should one begin phenomenology such that philosophizing can maintain itself in absolute radicality? For Heidegger, this means returning to the beginning point of phenomenological reflection, in the natural attitude and attempting to give a redescription of how human existence is first given. This is what Heidegger attempts to do in Paragraphs 12 and 13 of the *Prolegomena*, which in many ways are the most fascinating pages of the Preliminary Part of the lecture course, where despite giving a rather limited reading of the development of the personalistic attitude in *Ideas II*, he makes some more penetrating remarks on Dilthey's and Scheler's attempts to produce a personalistic psychology.[59]

The question motivating these Paragraphs is the following: how is human Dasein given in specifically personal experience (PGZ 162/HCT 117)? It is with the response to this question that Heidegger begins the existential analytic of Dasein in Paragraph 9 of *Sein und Zeit* (SZ 41–42). In this sense, the beginning of Heidegger's philosophical project is not only methodologically dependent upon Husserlian phenomenology, but can be seen specifically as a radicalized extension of the phenomenology of the person in Dilthey, Scheler and the later Husserl. As Heidegger rather gnomically remarks at the beginning of the Main Part of the *Prolegomena*, "There is an *intrinsic material connection* [*innerlicher sachlicher Zusammenhang*] between what we treated in the Introduction [i.e. the Preliminary Part, s.c.] and what we now take as our theme" (PGZ 192/HCT 141–42). To put this in terms that Heidegger would doubtless have refused, the First Division of *Sein und Zeit* attempts to transform the natural attitude with which phenomenology begins. Access to the beginning point of Heidegger's existential analytic is achieved by a transformation in our understanding of the natural attitude, what we might call a hermeneutic redescription of this moment of facticity.[60]

Let me pause and try to clarify this point. Phenomenology begins in the natural attitude, as a description of our pre-theoretical immersion in the familiar, everyday, environing world, as the reality of our intentional lives.[61] This leads

40

Heidegger to raise the question: "To what extent is the being of the intentional experienced and determined in this starting position?" (PGZ 152/HCT 111). That is, is there a moment when the question of the being of the intentional is raised by phenomenology if only to be subsequently discarded?

This moment is that described by Husserl as the general thesis of the natural attitude. But, how is the natural attitude experienced in Husserlian phenomenology? As Heidegger puts it, "what being is attributed to it?" (PGZ 153/HCT 111). Heidegger claims that the reality of the natural attitude is experienced as "real occurrences" which are "objectively on hand" ("*objektiv vorhanden*"). That is, in the Husserlian natural attitude, things are experienced in the mode of *Vorhandenheit*, as objects (*Gegenstände*) available to a theoretical inspection by consciousness, as things standing over against (*gegen*) a subject. But that is not all. Not only are things experienced in the mode of *Vorhandenheit* as objects, but this is also the determination of the being of the person intentionally relating to things. Thus, the being for whom the world appears in its reality as something on hand to a theoretical regard is also fixed as something real and on hand, as an entity objectified into an ego. Such is the *Boden* upon which the impoverished world of naturalism erects its structures.

Thus, Heidegger's claim is that the Husserlian understanding of the natural attitude presupposes an understanding of both things and persons that is part of an ontology of *Vorhandenheit*, the present-at-hand, to which Heidegger will oppose, in the opening chapters of *Sein und Zeit*, an ontology of things based in the category of *Zuhandenheit*, or handiness, and a fundamental ontology of persons rooted in the facticity of *Existenz*.

But is the natural attitude natural? Is it even an attitude? Heidegger seems to respond with a double negative. Let me take up the first question: is the natural attitude natural? The natural attitude is unnatural because it presupposes a particular theoretical orientation borrowed from tradition and not taken from the things themselves. That is, the natural attitude is a theoretical attitude, and insofar as it is theoretical the philosophical obligation of the phenomenologist is to work against it in order to be true to the maxim "to the things themselves". If our access to things were not blocked by the theoreticist prejudice of the tradition, then the maxim "to the things themselves" would have no meaning, for we would already be with those things.

This is a point fascinatingly amplified by Levinas in the conclusion to his 1930 Doctoral Thesis *The Theory of Intuition in Husserl's Phenomenology*, a work utterly pervaded by the climate of the early Heidegger, and where Levinas completely accepts the necessity for an ontological critique of phenomenology and claims that the natural attitude is fatally framed by the presuppositions of a representationalist epistemology.[62] Levinas argues that Husserlian phenomenology is theoreticist and intellectualist and thereby overlooks the historical situatedness of the human being, which is a claim that Levinas obviously made in ignorance of the *Krisis* manuscripts. He writes: "*Par conséquent, malgré le caractère révolutionnaire de la réduction phénoménologique, la révolution qu'elle accomplit est, dans la philosophie de Husserl, possible de par la nature de l'attitude naturelle, dans la mesure où celle-ci est théorique.*"[63] ("Consequently, despite the revolutionary character of the phenomenological reduction, the revolution that it accomplishes is, in Husserl's philosophy, possible because of the nature of the natural attitude, to the extent that the natural attitude is theoretical.")

Of course, the dramatic irony of Levinas's remarks in relation to his later critique of the fundamentality of ontology must be noted, and I have explored this elsewhere.[64] But, crucially, Levinas's later claim that ethics and not ontology is first philosophy continually presupposes the Heideggerian critique of Husserl. This is why, in the introduction to *De l'existence à l'existant* in 1947, he claims that not only are his reflections commanded by the need to leave the climate of Heidegger's philosophy, but—more importantly—that one cannot leave that climate for a philosophy that would be pre-Heideggerian: "*...on ne saurait pas en sortir vers une philosophie qu'on pourrait qualifier de pré-heideggerienne*".[65] Interestingly, given the hysteria that broke out in France because of "the Heidegger affair", one finds a similarly measured tone with regard to Heidegger in a paper given some 40 years after *De l'existence à l'existant*, at the height of the affair in 1987.[66] The Heideggerian paradigm shift in twentieth century philosophy is as important a turning point as Hegel's for the nineteenth century, which is a point that even Habermas begrudgingly concedes.[67] Everything turns here on Levinas's word "climate", which I would choose to view as a translation of *ethos*, and, of course, it is with that word that all the problems with Heidegger begin.

Turning to the second question, if the natural attitude is not natural, then, secondly, it is also not an attitude. The human being's "natural" manner of experiencing the world is not an *Einstellung*, something I put myself into (*einstellen*) in the same way as I might put a car in the garage, a book on the shelf or my pet hamster in the refrigerator. Why? Because I always already find myself (*ich befinde mich*) in the world; I am always already practically disposed in a world that is familiar and handy, a world in which we are immersed and with which we are fascinated. Thus, adopting an attitude towards experience is already to look at things from the standpoint of reflection, in an act by which we consider life, but no longer live it.

Thus, the Heideggerian beginning point for the question of the being of the intentional is already distorted by the Husserlian description of that beginning point with the thesis of the natural attitude. That is, it is the wrong description of the right beginning point. The natural attitude, with its theoreticist, intellectualist, *vorhanden* understanding of reality and consciousness is an unphenomenological distortion of the human being's primary practical and personal access to the world. In this regard, Heidegger's *Sein und Zeit* can be seen as attempting to give an interpretative clarification of what is first given in personal experience, a hermeneutic redescription of the natural attitude.

Of course, the meta-question that should be raised here is whether Heidegger is justified in his critique of the natural attitude in Husserl. Even if it is granted that he gives a plausible interpretation of the natural attitude in *Ideas I*, then is this valid for Husserl's later work? In this regard, simply as a counterbalance to Heidegger's claims, we might consider Merleau-Ponty's claims about the natural attitude in his stunning late essay, "Le philosophe et son ombre".[68] Although the avowed hermeneutic strategy employed by Merleau-Ponty in this essay is Heideggerian, attempting to locate the unthought in Husserl's texts, the whole essay can be read as a problematization of Heidegger's portrayal of transcendental phenomenology, based on a reading of *Ideas II*.[69]

Of course, the unpublished manuscript of the latter text was lying on Heidegger's desk in 1925 and he even refers obliquely to it in an early footnote to *Sein und Zeit* (SZ 47). For Merleau-Ponty, "It is the natural attitude that see-saws [*bascule*] in phenomenology". Or again: "When Husserl says that the reduction goes beyond the natural attitude, he immediately adds that this going beyond preserves 'the whole world of the natural attitude'."[70] That is to say, from *Ideas II* onwards, Husserl recognizes that the natural attitude contains a higher phenomenological truth that must be regained. To capture this truth, Husserl makes the distinction between the *naturalistic attitude*, or the theoretical rela-tion to *blosse Sachen* that defines the methodology of the natural sciences, and the *personalistic attitude*, which tries to capture the sense of life as it is lived in terms of what is first given in personal experience, what Merleau-Ponty calls "*notre proto-histoire*".[71]

So, the natural attitude only becomes the theoretical and intellectualist under-standing of things and persons when it is transformed into the naturalistic attitude. The task of a personalistic phenomenology, then, is one of trying to "unveil the pre-theoretical layer" (*dévoiler la couche pré-théorétique*) of human experience upon which the various idealizations of naturalism are based.[72] It is this obdurate yet almost intangible *Weltthesis* prior to all naturalistic theses that phenomenology has the job of elucidating, the mystery of an *Urglaube* or primal faith in the familiar that Merleau-Ponty tried to catch with the notion of *la foi per-ceptive*, the perceptual faith.

Merleau-Ponty, in a nice turn of phrase, describes the task of phenomenology as "unveiling the pre-theoretical layer" of human experience upon which the the-oretical attitude of the scientific conception of the world is based.[73] It is something like Merleau-Ponty's conception of phenomenology that I would like to defend here. On my understanding, it is a question of *doing* phenomenology in order to try and uncover the pre-theoretical layer of the experience of persons and things and to find a mode of felicitous description for this layer of experience with its own rigor and standards of validity. It is this obdurate yet almost intangi-ble dimension of pre-theoretical experience that phenomenology has the job of elucidating, the mystery of the familiar that Merleau-Ponty tried to articulate with the notion of the perceptual faith. That is, when I open my eyes and look around at the world, I have complete faith that it both exists and is richly meaningful. The problem is that this faith breaks down when I start to reflect on it and ask myself, "Well, how can I be certain that there is an external world for me when the evi-dence of my senses is not always completely reliable?" How does one regain the naivety of the perceptual faith when one has already attained the standpoint of reflection? Merleau-Ponty answers this problem with a notion of what he calls "hyper-reflection"; that is, phenomenology is a reflection upon what precedes reflection, the pre-theoretical substrate of experience. The point here is that access to the pre-theoretical level of human experience is not necessarily immediate for human beings like us who have attained the theoretical attitude of the sciences. Phenomenology therefore implies relearning to see the world in all its palpable and practical presence.

Doing phenomenology—neither scientism nor obscurantism

It is something like this conception of phenomenology that I want to defend. In a nutshell, I think this is why one should begin philosophizing with Heidegger rather than elsewhere. On my understanding, it is a question of *doing* phenomenology in order to try and uncover the pre-theoretical layer of the experience of persons and things and to find a mode of felicitous description for this layer of experience with its own standards of validity. For me, such a conception of phenomenology can be employed to avoid two pernicious tendencies in our current thinking: *scientism* and *obscurantism*.

Let me begin with scientism. Scientism rests on the fallacious claim that the theoretical or natural scientific way of viewing things, what Heidegger calls *Vorhandenheit*, provides the primary and most significant access to ourselves and our world, and that the methodology of the natural sciences provides the best form of explanation for all phenomena *überhaupt*. Heidegger shows that the scientific conception of the world, what Carnap and Neurath called the *wissenschaftliche Weltauffassung*, is derivative or parasitic upon a prior practical view of the world as *zuhanden*, that is, the environing world that is closest, most familiar, and most meaningful to us, the world that is always already colored by our cognitive, ethical and aesthetic values. That is to say, scientism, or what Husserl calls objectivism, overlooks the phenomenon of the *life-world* as the enabling condition for scientific practice. In the *Crisis of the European Sciences*, Husserl describes the life-world in the following way:

> It belongs to what is taken for granted, prior to all scientific thought and all philosophical questioning, that the world is—always is in advance—and that every correction of an opinion, whether an experiential or other opinion, presupposes the already existing world, namely as a horizon of what in the given case is indubitably valid as existing... Objective science, too, asks questions only on the ground of the world's existing in advance through pre-scientific life.[74]

The critique of scientism, at least within phenomenology, does not seek to refute or negate the results of scientific research in the name of some mystical apprehension of the unity of man and nature, which is a risk in some of the slightly ecstatical pronouncements of the later Merleau-Ponty; rather, it simply insists that science does not provide the primary or most significant access to a sense of ourselves and the world. Anti-scientism does not at all entail an anti-scientific attitude, and nor does it mean that "science does not think", which is a remark of Heidegger's that has caused more problems than it has solved. What is required here is what Heidegger called, in a much-overlooked late remark in *Sein und Zeit*, "*an existential conception of science*" ("*einen **existenzialen Begriff der Wissenschaft**", SZ 357) that would show how the practices of the natural sciences arise out of life-world practices, and that the latter are not simply reducible to the former.[75]

Moving to more contemporary philosophical concerns, it is at least arguable that such a position is approached by John McDowell in his highly influential *Mind and World*.[76] McDowell borrows Aristotle's notion of second nature and Hegel's notion of *Bildung* in order to try and escape the traditional predicament of philosophy, namely the epistemological subject–object construal of how to relate thought to things and mind to world and, in particular, the naturalistic version of that construal in someone like Quine. McDowell seeks to avoid the Scylla of "bald naturalism" (the reduction of reason to nature) without falling into the Charybdis of "rampant Platonism" (the idealist separation of reason from nature). What is so interesting about McDowell for my purposes is that the view he advances, what he calls "naturalized Platonism", implicitly borrows at least four Heideggerian themes (via Gadamer's account of them in *Truth and Method*—a choice that is itself revealing and question-begging):

- the unintelligibility of skepticism, which recalls the argument of Paragraph 44 of *Sein und Zeit* (p. 113);
- the attempt to construe experience as "openness to the world" which recalls Heidegger's notions of disclosure and the clearing (*die Lichtung des Seins*);
- the idea that human life in the world is structured environmentally, which recalls Heidegger's idea that *Welt* is first and foremost an *Umwelt* (p. 115);
- the claim that language is the repository of tradition, which recalls Heidegger's ideas about historicity and heritage (p. 126).

Thus, the attempt to avoid the traditional predicament of philosophy, and the baldly naturalistic construal of that predicament, leads someone like McDowell to the adoption of a number of leading Heideggerian motifs. I don't think we should therefore fall to our knees in worship of McDowell. However, what is fascinating is the way in which he correctly diagnoses a deep predicament in traditional Anglo-American philosophy and questions that predicament such that the point at which someone like Heidegger begins philosophizing might begin to be intelligible to Anglo-American philosophers.

Also interesting in this regard is Robert Brandom's rather Hegelian reconstruction of the argument of *Sein und Zeit*. Brandom tries to show how the

Heideggerian claim that the present-at-hand arises out of the ready-to-hand—that is, how knowing is a founded mode of being-in-the-world—implies a social ontology where the condition of possibility for the scientific, criterial identification of entities (Quinean ontology) arises out of a shared communicative praxis based on a mutual recognition of shared norms (fundamental ontology). Such is the position that Brandom describes as Heidegger's "ontological pragmatism"; that is, it is a question of acknowledging and describing the social genesis of the categories and criteria with which the world is described, "...fundamental ontology...is the study of the nature of social being—social practices and practitioners".[77]

Let me develop this point a little further with reference to Heidegger's notion of phenomenology as a pre-science (*Vor-wissenschaft*). Although one can find this idea in Heidegger as early as his 1919 lecture course *The Idea of Philosophy and the Problem of Worldviews* (*Die Idee der Philosophie und das Weltanschauungsproblem*),[78] it is also prominently discussed in the 1924 lecture *The Concept of Time* (*Der Begriff der Zeit*), which Gadamer famously and rightly described as the *Ur-form* of *Sein und Zeit*.[79] In the latter lecture, Heidegger describes his reflections as belonging neither to theology nor to philosophy, but rather to a pre-science (*Vor-Wissenschaft*), that would be a hermeneutics of the factical conditions of possibility for scientific research, i.e. their social genesis in life-world practices. In what I shall generously assume is an attempt at humor on Heidegger's part, he describes this pre-science as the police force (*Polizeidienst*) at the procession of the sciences, conducting an occasional house search of the ancients and checking whether scientific research is indeed close to its matter (*bei ihrer Sache*), and hence phenomenological, or whether science is working with a traditional or handed down (*überlieferten*) knowledge of its *Sache*. (One imagines the mass arrest and detention of whole crowds of naturalists by such a phenomenological police force, with summary beatings, torture and execution for the worst scientistic offenders.)

In the *Prolegomena*, this phenomenological policing is called—and it is a phrase retained in Paragraph 3 of *Sein und Zeit*—a *productive logic* (SZ 10; PGZ 2/HCT 2). That is, it is a pre-scientific disclosure of the life-world that leaps ahead (*vorausspringt*) and lays the ground for the sciences.[80] What Heidegger would seem to mean here is that, unlike the empiricist or Lockeian conception of the philosopher as an underlaborer to science, a productive phenomenological logic—which for Heidegger corresponds to the original logic of Plato and Aristotle—leaps ahead of the sciences by showing their basis in a fundamental ontology of persons, things and world, the pre-theoretical layer of experience spoken of above. What I have called "a phenomenological pre-science" or "an existential conception of science" does not dispute or refute the work of the sciences. On my understanding, it shows three things:

- that the theoretical attitude of the sciences finds its condition of possibility in our various life-world practices;

- that such practices require hermeneutical clarification and not causal hypotheses or causal-sounding explanations;
- that the formal a priori structure of persons, things and world can be deduced from that hermeneutic clarification, which is what Heidegger attempts to do with his various "existentials". The latter are what Heidegger calls "formal indications", a key term in Heidegger's early work.

What phenomenology provides is a clarifying redescription of persons, things and the world we inhabit. As such, phenomenology does not produce any great discoveries, but rather gives us a series of reminders of matters with which we were acquainted, but which become covered up when we assume the theoretical attitude of the natural sciences. Phenomenology provides what we might call "everyday anamnesis", a recollection of the collection of background practices and routines that make up the delicate web of ordinary life.

Allow me a final word on obscurantism. It is important to point out that such a phenomenological anti-scientism *can* lead to an anti-scientific obscurantism, which in many ways is the inverted or perverted counter-concept to scientism, but it *need* not do so if we are careful enough to engage in a little intellectual policing. Obscurantism might here be defined as the rejection of the causal explanations offered by natural science by referring them to an alternative causal story, that is somehow of a higher order, but essentially occult. That is, obscurantism is the replacement of a scientific form of explanation, which is believed to be scientistic, with a counter-scientific, mysterious, but still causal explanation. For example, the awful destruction wreaked by the tsunami in the Indian Ocean in 2004 was not caused by plate tectonics but by God's anger at our sinfulness.

As a cultural phenomenon, this is something that can be observed in every episode of *The X-Files*, where two causal hypotheses are offered, one scientific, the other occult, and where the former is always proved wrong and the latter is right, but in some way that still leaves us perplexed. Now, as a cultural distraction, arguably this does little harm, but elsewhere its effects can be more pernicious. Familiar candidates for obscurantist explanation are the will of God, the ubiquity of alien intelligence, the action of the stars on human behavior, or whatever. Less obvious, but arguably equally pernicious, candidates are the drives in Freud, Jung's archetypes, the real in Lacan, power in Foucault, différance in Derrida, the trace of God in Levinas, or, indeed, the epochal withdrawal of being in and as history in the later Heidegger. This list might be extended. I am broadly suspicious of what I call "one big thing" approaches to philosophy, where all phenomena are explained with reference to one big thing that is behind the scenes pulling the strings. Although this is not the place to develop this thought, I am in favor of many slightly smaller things that require different interpretative frameworks and approaches.

In my view, what we can still learn from phenomenology is that when it comes to our primary and most significant access to persons and things, what we might call our entire stock of tacit, background know-how about the social world, we do not require causal scientific explanations, or pseudo-scientific hypotheses in rela-tion to obscure causes, but what I am tempted to call, thinking of Wittgenstein, *clarificatory remarks*. For example: "The aspects of things that are most impor-tant for us are hidden because of their simplicity and familiarity. (One is unable to notice something—because it is always before one's eyes.)"[81] Clarificatory remarks bring into view features of our everyday life that were hidden but self-evident, and hidden because they were self-evident. They make these phenomena more perspicuous, change the aspect under which they are seen, and give to mat-ters a new and surprising overview. In this sense, phenomenology is a reordering of what was tacitly known but went unnoticed; it permits us to relearn how to look at the world. Of course, viewing Heidegger's work in this way does not sound as exciting as talking about the epochal donation of being in its withdrawal or what-ever, but perhaps that sort of excitement is something we are best off without.

It should be clear from what I have been saying that I am attempting a mini-pathology of the contemporary philosophical scene, which is meant to comment on—and maybe curb—the worst excesses of both Continental and analytic phi-losophy. On the one hand, there is a risk of obscurantism in some Continental philosophy, where social phenomena are explicated with reference to forces, enti-ties and categories so vast and vague as to explain everything and nothing at all. For example, a phenomenon like the Internet, cell phones or speed dating might be seen as further evidence of Heidegger's thesis on what he calls the *Gestell*, the enframing attitude that prevails in the technological world, and thereby tributary to the forgetfulness of being. As such, everyday phenomena are seemingly explained with reference to causal-sounding agencies that function something like the gods in ancient mythology. Any aspect of personal and public life might be seen as evidence of the disciplinary matrices of power, the disintegration of the "Big Other" and the trauma of the real, the multiple becomings of the body with-out organs, or whatever. Where such obscurantist tendencies exist, then the therapy has to be demystification, or what Jack Caputo calls with respect to Heidegger "demythologization",[82] that is, a critique of this kind of talk and per-haps also some suggestions as to why we engage in it in the first place.

But, on the other side of my mini-pathology, there is the risk of a chronic sci-entism is some areas of analytic philosophy. As Frank Cioffi wittily remarks, if we can imagine a philosophical paper with the title "Qualia and Materialism: Closing the Explanatory Gap", then why not papers with titles such as "The Big Bang and Me: Closing the Explanatory Gap" or "Natural Selection and Me: Closing the Explanatory Gap"?[83] The assumption of such scientistic approaches is that there is a gap that can be closed through a better empirical explanation. I have argued elsewhere that there is a felt gap here—the gap between knowledge and wisdom—that cannot be closed through empirical inquiry.[84] That is, to put it bluntly, the question of the meaning and value of life in the world is not reducible to empirical inquiry. In philosophy, but also more generally in cultural life, we need to clip the wings of both scientism and obscurantism and thereby avoid what is worst in both Continental and analytic philosophy. That is, we need to avoid the error of believing that we can resolve through causal or causal-sounding explanation what demands phenomenological clarification. Of course, this is much easier said than done, but at least we could make a start.

Of course, the distinction between scientism and obscurantism is not as neat as I might appear to have made it. First, obscurantism might not be one thing, as I seem to suggest. Namely, there is indeed the obscurantism based on faith in some sort of numinous enigma, whether Zeus, Yahweh or the death drive. We might call this "obscure obscurantism". But there are other obscurantisms that do not believe themselves to be obscure, but perfectly self-evident or even scientifically provable: "Doctor, can't you see that my sleeplessness and aggression is caused by the fact that I was abducted by aliens when I was camping last summer?"; or "Just one more year of research and I will finally have proved that matter is the product of divine effusions"; or, in a more everyday psychotic way, which is utterly disturbing, "Doctor, don't you realize that the pain in my liver is caused by my dead mother's anger towards me". And of course there are scientisms that are taken on faith and are thus the equivalent of obscurantism. For example, I might believe that all mental states can be reduced to evolutionary dispositions or neural firings without knowing how or why, it just feels right. We might call this an "obscure scientism" or whatever. Let's just say that there is a pressing need for a more detailed taxonomy of the scientism/obscurantism distinction.

In order to avoid the intellectual and cultural cul-de-sac of the stand-off between obscurantism and scientism, I think we need to remind ourselves of a classical distinction, first coined by Max Weber, between explanation and clarification, that is, between causal or causal-sounding hypotheses and demands for elucidation, interpretation or whatever. Roughly and readily, Weber's claim is that natural phenomena require causal explanation, while social phenomena require clarification by giving reasons or offering possible motives as to why something is the way it is. One of the jobs of philosophy is to remind us that we urgently need to make this distinction, and that if we don't then we may end up falling into either scientism, obscurantism, or the tempting twilight zone of the *X-Files* complex. It has been my contention in these lectures that the best way of making sure we make this distinction is through a version of phenomenology.[85]

Conclusion

Let me summarize where we have got to in the account of Heidegger and Husserl. Heidegger's double gesture with regard to Husserl shows the duplicity of a philosophical inheritance—and perhaps duplicity is the defining feature of philosophical inheritance, from Aristotle's metaphysical misunderstanding of Plato onwards. But, for Heidegger, the Preliminary Part of the *Prolegomena* shows the necessity for the move from phenomenology to ontology. That is to say, in Heidegger's hands, phenomenological method is led irresistibly to deal with ontological questions: the question of the being of intentional consciousness, the question of the being of the person (for Husserl and Scheler) or *Dasein* (for Heidegger) for whom intentionality is its essential mode of comportment towards things, and ultimately the question of being itself.

Of course, these questions are pushed further by Heidegger's work after *Sein und Zeit*, and lead to the dropping of the title "fundamental ontology", which risked being misconceived metaphysically as a foundational ontology. But what is interesting about the *Prolegomena* is that one can see how the necessity for Heidegger's existential analytic of Dasein as an access to the question of being itself arises out of a double reading of Husserl. The existential analytic of Dasein is the concretization or de-formalization (*Entformalisierung*) of the formal-methodological conception of phenomenology (SZ 35). However, everything hangs on the passage from the formal to the concrete. That is, does the force of the concrete analyses of *Sein und Zeit* distort, work against, or perhaps even exceed the formal-methodological conditions of possibility for that analysis? Such a question can only begin to be answered through a reading of *Sein und Zeit*.

Notes

1 The following text is based on notes for a series of lectures on Heidegger given at the New School for Social Research in Spring 2005. The ideas had been working themselves out over the previous decade while teaching at the University of Essex. A portion of the concluding part of text was published in 2000 as "Heidegger for Beginners", in *Appropriating Heidegger*, eds. J. Faulconer and M. Wrathall (Cambridge: Cambridge University Press), pp. 101–18. I am deeply grateful to students and colleagues for listening and responding to what I had to say.

2 Deleuze, G. and Guattari, F. (1994) *What is Philosophy?* Trans. G. Burchell and H. Tomlinson, London: Verso, p. 108.

3 Heidegger, M. (1988) *Zur Sache des Denkens*, Tübingen: Niemeyer, 3rd edn., p. 48.

4 von Herrmann, F. W. (1981) *Der Begriff der Phänomenologie bei Heidegger und Husserl*, Frankfurt am Main: Klostermann, p. 8.

5 Held, K. (1988) "Heidegger und das Prinzip der Phänomenologie", in *Heidegger und die praktische Philosophie*, eds. A. Gethman-Siefert and O. Pöggeler, Frankfurt am Main: Suhrkamp, p. 113. See also Held, K. (1999) "On the Way to a Phenomenology of World", *Journal of the British Society for Phenomenology*, 30(1): pp. 3–17.

6 See Taminiaux, J. (1985) "Heidegger and Husserl's *Logical Investigations*: in Remembrance of Heidegger's Last Seminar (Zähringen, 1973)", in *Dialectic and Difference*, New Jersey: Humanities Press, pp. 91–114. See also: "D'une idée de la phénoménologie à l'autre", in *Lectures de l'ontologie fondamentale*, 1989, Grenoble: Millon, pp. 19–88.

7 *Zur Sache des Denkens*, op. cit., p. 90.

8 I refer here to William J. Richardson's *Heidegger. Through Phenomenology to Thought*, 1963, The Hague: Martinus Nijhoff.

9 The extent of Heidegger's debt to Dilthey has become increasingly apparent with the discovery in 1989 of the so-called "Aristoteles-Einleitung" written by Heidegger in three weeks and sent to Paul Natorp in connection with Heidegger's candidature for a position in Marburg in 1922. Much has been made of the link to Dilthey by Theodore Kisiel in his *The Genesis of Being and Time* (1993, Berkeley: University of California Press, pp. 315–61), in particular with connection to lectures that Heidegger gave in Kassel in April 1925, just before the beginning of the lecture course on Husserl.

10 Heidegger, M. (1982) *The Basic Problems of Phenomenology*. Trans. A. Hofstadter, Bloomington: Indiana University Press, p. 65.

11 Taminiaux, J., *Lectures de l'ontologie fondamentale*, op. cit., p. 59. One also finds this line of thought on consciousness in late essays by William James, for example: (1996) "Does 'Consciousness' Exist?", in *Essays in Radical Empiricism*, Lincoln: University of Nebraska Press, pp. 1–38.

12 For the discussion of Angelus Silesius's "Rose without why", see Heidegger, M. (1991) *The Principle of Reason*. Trans. R. Lilly, Bloomington: Indiana University Press, pp. 35–40.

13 James, W. (1996) *Some Problems of Philosophy. The Beginning of an Introduction to Philosophy*. Lincoln and London: University of Nebraska Press, p. 107–8.

14 Aristotle, *Topics*, Book 1, Chapter 9.

15 According to Kisiel, this distinction postdates the *Prolegomena* in 1925 and was introduced only with the drafting of *Sein und Zeit* in March 1926. See *The Genesis of Being and Time*, op. cit., p. 489.

16 Kant, I. *Critique of Pure Reason*, B105–6.

17 However, it is precisely in terms of such a return to a pre-modern notion of intellectual intuition that Richard Cobb-Stevens interprets Heidegger's notion of categorial intuition. See his (1990) *Husserl and Analytic Philosophy*, Dordrecht: Kluwer. See also his (1994) "The Beginnings of Phenomenology: Husserl and his predecessors", in *Encyclopaedia of Western Philosophy, Vol. 8 Twentieth Century Continental Philosophy*, ed. Richard Kearney, London: Routledge, pp. 5–37. Cobb-Stevens begins his article, "Edmund Husserl was the founder of phenomenology, one of the principal movements of twentieth century philosophy. His principal contribution was his development of the concept of intentionality. He reasserted and revitalized the premodern thesis that our cognitional acts are intentional, i.e. that they reach out beyond sensa to things in the world", p. 5.

18 Gadamer uses this expression, of course, not in relation to Kant's epistemology, but his aesthetics. See (1979) *Truth and Method*. Trans. W. Glen-Doepel, London: Sheed and Ward, p. 39.

19 Husserl, E. (1970) *Logical Investigations*. Trans. J.N. Findlay, London: Routledge, pp. 832–4; the quotation is from p. 833.

20 See PGZ 168/HCT 121; and Taminiaux, *Lectures de l'ontologie fondamentale*, op. cit., p. 88.

21 *Logical Investigations*, op. cit., p. 782.

22 Ibid., p. 782.

23 Ibid., p. 738.

24 Ibid., pp. 579–80.

25 Ibid., p. 611.

26 Ibid., pp. 783–4.

27 See Heidegger, M. (2003) *Four Seminars*. Trans. A. Mitchell and F. Raffoul, Bloomington: Indiana University Press, p. 67.

28 *Logical Investigations*, op. cit., p. 784.

29 Cobb-Stevens, *Husserl and Analytic Philosophy*, op. cit., pp. 152–4.

30 *Logical Investigations*, op. cit., pp. 799–800.

31 *Husserl and Analytic Philosophy*, op. cit., p. 23 *et passim*.

32 Bell, D. (1990) *Husserl*, London and New York: Routledge, p. 111. Bell attempts to reconstruct categorial intuition with the help of Wittgenstein's notion of aspect seeing, which I have incorporated into the argument on p. 26.

33 See *Logical Investigations*, op. cit., pp. 817–18.

34 Wittgenstein, L. (1958) *Philosophical Investigations*. Trans. G. E. M. Anscombe, Oxford: Blackwell, Part II, no. xi.

35 See Bernet, R. (1988) "Perception, Categorial Intuition and Truth in Husserl's Sixth
    Logical Investigation", in *The Collegium Phaenomenologicum: The First Ten Years*.
    Ed. J. C. Sallis, G. C. Moneta and J. Taminiaux, Dordrecht: Kluwer, pp. 33–45. See
    esp. pp. 36–7.

36 Husserl, E. *Logical Investigations*, op. cit., p. 833, my emphasis.

37 Husserl, E. (1990) *The Idea of Phenomenology*. Trans. W. Alston and G. Nakhnikian,
    Dordrecht: Kluwer, pp. 7 and 41–2.

38 See also Heidegger's intriguing and untranslatable remark from Paragraph 18 of *Sein
    und Zeit*, "*Das auf Bewandtnis hin freigebende Je-schon-haben-bewenden-lassen ist ein
    a priorisches Perfekt, das die Seinsart des Daseins selbst carakterisiert*" (SZ 85).
    Bewandtnis can be rendered as involvement or situation and is Heidegger's term for the
    being of the *zuhanden*. What this means is that the ready-to-hand world is a totality of
    referential relations between things (glasses to pen, pen to paper, paper to table, table to
    room) that Heidegger calls *das Zeug*; a practical world that is familiar and significant to
    us and with which we are *benommen*, taken up with or fascinated. Heidegger's claim is
    that this structure of involvement is given a priori is *Perfekt* in the German grammatical
    tense that refers to a past that is *immer schon da*. Thus the a priori structures by virtue
    of which the world is understood are temporal, they are *past* structures of world under-
    standing into which we are always already thrown. In one of Heidegger's marginal
    comments to *Sein und Zeit* printed in the *Gesamtausgabe* edition, he intriguingly links
    this notion of the a priori perfect tense to Aristotle's understanding of being in the
    Metaphysics, and to Kant's doctrine of the schematism (SZ 441–42).

39 See also Heidegger's discussion of the a priori structure of care as the "earlier" in con-
    nection with the concept of reality (SZ 206).

40 *Basic Problems of Phenomenology*, op. cit., pp. 324–7.

41 Heidegger, M. (1999) *Ontology—The Hermeneutics of Facticity*. Trans. J. van Buren,
    Bloomington and Indianapolis: Indiana University Press, pp. 6–16.

42 See Dreyfus, H. (1991) *Being-in-the-World*. Cambridge, MA: MIT, pp. 30–3. See also
    Heidegger's reference, between scare quotes, to "descriptive phenomenology" as a tau-
    tology (SZ 35).

43 See, for example, *Zur Sache des Denkens*, op. cit., p. 67.

44 See "On the way to a phenomenology of world", op. cit., pp. 3–17.

45 For a very good example of such an approach, see Bernasconi, R. (1984) *The Question
    of Language in Heidegger's History of Being*. Atlantic Highlands, NJ: Humanities
    Press.

46 Heidegger, M. (1997) *Plato's Sophist*. Trans. R. Rojcewicz and A. Schuwer,
    Bloomington: Indiana University Press, pp. 5–7.

47 von Herrman, F. W. (1987) *Hermeneutische Phänomenologie des Daseins*, Band 1,
    Frankfurt am Main: Klostermann.

48 Dreyfus, H. *Being-in-the-World*, op. cit., pp. 30–3.

49 Ibid., p. 30.

50 Courtine, J. F. (1990) *Heidegger et la phénoménologie*. Paris: Vrin. See especially the
    essay "La cause de le phénoménologie", op. cit., pp. 161–85.

51 Heidegger, M. *Plato's Sophist*, op. cit., p. 413.

52 See Courtine, J. F. "Phénoménologie/ou tautologie", in *Heidegger et la phénoménolo-
    gie*, op. cit., pp. 381–405. (Translated in (1993) *Reading Heidegger. Commemorations*.
    Ed. J. Sallis, Bloomington: Indiana University Press, pp. 241–57.)

53 Heidegger discusses Rimbaud in the seminar that follows *Zeit und Sein*, in *Zur Sache
    des Denkens*, op. cit., pp. 42–3.

54 Heidegger, M. (1954) *Aus der Erfahrung des Denkens*, Pfullingen: Neske, p. 27.

55 Husserl, E. (1970) "The Origin of Geometry", in *The Crisis of European Sciences and
    Transcendental Phenomenology*. Trans. D. Carr, Evanston: Northwestern University
    Press, p. 361.

56 Wittgenstein, L. (1958) *Philosophical Investigations*. Trans. G.E.M. Anscombe, 2nd edn, Oxford: Blackwell, No. 90: "We feel as if we had to see *through* appearances. Our investigation does not direct itself to appearances, but, one might say, to the '*possibilities*' of appearances."

57 On this point, see the opening pages of Reiner Schürmann's lectures on *Sein und Zeit* in this volume.

58 Husserl, E. (1976) *Ideas*. Trans. W. R. Boyce Gibson, New York: Humanities Press, p. 151.

59 For a thorough account of the influence of Dilthey on Heidegger in his critique of Husserl, see van Buren, J. (1994) *The Young Heidegger*, Bloomington: Indiana University Press. See esp. the discussion of Husserl, pp. 203–19. Van Buren very usefully analyses Heidegger's 1925 lectures on Dilthey, held in Kassel on 16–21 April, under the title "Wilhelm Dilthey's Forschungsarbeit und der Kampf um eine historische Weltanschauung".

60 This is a point well discussed by Barbara Merker in her interesting book (1988), *Selbsttäuschung und Selbsterkenntnis. Zu Heideggers Transformation der Phänomenologie Husserls* (Frankfurt am Main: Suhrkamp). See esp. pp. 7–9 and 78–80. Merker, reading *Sein und Zeit* as a quasi-Christian *Konversionsgeschichte*, shows how Heidegger replaces the natural attitude with the realm of inauthenticity:

> *Die 'narzißtischen' Projektionen des phänomenologischen Theoretikers verhindern demnach eine adäquate Analyse des alltäglichen Besorgens, der Gegenstände, mit denen es umgeht, wie der Welt, in der es sich bewegt. Nur wenn es gelingt, dieses theoretisches Vorurteil zu vermeiden, wird eine adäquate Beschreibung der alltäglichen Existenzweise möglich, die Heidegger an die Stelle der 'natürlichen Einstellung' Husserls setzt. (p. 79)*

61 Husserl, E. *Ideas*, op. cit. pp. 105–6.

62 Levinas, E. (1989) *La théorie de l'intuition dans la phénoménologie de Husserl*, 6th edn, Paris: Vrin, pp. 219–23.

63 Ibid., p. 222.

64 Critchley, S. (1999) "Post-Deconstructive Subjectivity?", in *Ethics–Politics–Subjectivity. Essays on Derrida, Levinas and Contemporary French Thought*, Verso: London and New York, pp. 51–81.

65 Levinas, E. (1986) *De l'existence à l'existant*, 2nd edn, Paris: Vrin, p. 19.

66 See "Mourir pour", in Levinas, E. (1988) *Heidegger. Questions ouvertes*. Ed. J Derrida, Paris: Osiris, pp. 255–64.

67 "From today's standpoint, Heidegger's new beginning still presents probably the most profound turning point in German philosophy since Hegel." See Habermas, J. (1992) "Work and *Weltanschuung*: The Heidegger Controversy from a German Perspective", in *Heidegger: A Critical Reader*. Ed. H. Dreyfus and H. Hall, Oxford: Blackwell, p. 188.

68 See Merleau-Ponty, M. (1960) *Éloge de la philosophie et autres essais* (Paris: Gallimard), pp. 241–87. Translated in (1964) *Signs* Trans. R. McCleary, Evanston: Northwestern University Press, pp. 159–81.

69 Husserl, E. (1989) *Ideas, Book 2, Studies in the Phenomenology of Constitution*. Trans. R. Rojcewicz and A. Schuwer, Dordrecht: Kluwer. For the reference to Heidegger's notion of *das Ungedachte*, see "Le philosophe et son ombre", op. cit., p. 243.

70 Ibid., p. 252 and p. 248.

71 Ibid., p. 286.

72 Ibid., p. 253.

73 Merleau-Ponty, M. (1964) "The Philosopher and his Shadow", in *Signs*. Evanston: Northwestern University Press, p. 253.

74 Husserl, E. *The Crisis of European Sciences and Transcendental Phenomenology*, op. cit., p. 110.

75 For a Heideggerian approach to science that argues for a "robust realism" where science gives us access to things independently of our everyday practices, see Dreyfus, H. and Spinosa, C. (1999) "Coping with Things-in-themselves: A Practice-Based Phenomenological Argument for Realism", *Inquiry*, 42(1): pp. 49–78. See also the five responses to this paper in the same issue of the journal, and Dreyfus's and Spinosa's response to their critics in *Inquiry*, 42(2): pp. 177–94. For a more "deflationary realist" account of Heidegger and science, see Rouse, J. (1987) *Knowledge and Power*, Ithaca, NY: Cornell University Press. See also Rouse's very helpful article, (1998) "Heideggerian Philosophy of Science", in the *Routledge Encyclopedia of Philosophy*. Ed. E. Craig, London and New York: Routledge, Vol. IV, pp. 323–7.

76 McDowell, J. (1994) *Mind and World*, Cambridge, MA, Harvard University Press. Subsequent page references given in the text.

77 See Brandom's "Heidegger's Categories in *Being and Time*", in *Heidegger: A Critical Reader*, op. cit., pp. 45–64. See esp. pp. 53–5 and p. 62. See also in this regard, Okrent, M. (1988) *Heidegger's Pragmatism*, Ithaca, NY: Cornell University Press.

78 In Heidegger, M. (1987) *Zur Bestimmung der Philosophie, Gesamtausgabe*. Frankfurt am Main: Klostermann, vols 56–7.

79 Heidegger, M. (1989) *Der Begriff der Zeit*. Tübingen: Niemeyer; *The Concept of Time* (1992). Trans. W. McNeill, Oxford: Blackwell.

80 Heidegger also speaks of *Vorausspringen* as stepping in for the other as the positive mode of solicitude (*Fürsorge*) in *Sein und Zeit* (SZ 122).

81 *Philosophical Investigations*, op. cit., no. 129.

82 Caputo, J. D. (1993) *Demythologizing Heidegger*, Bloomington: Indiana University Press.

83 Cioffi, F. (1998) *Wittgenstein on Freud and Frazer*, Cambridge: Cambridge University Press, p. 302.

84 See Critchley, S. (2001) *Continental Philosophy: A Very Short Introduction*, Oxford: Oxford University Press, chs 1 and 6, pp. 1–11 and 90–110.

85 The phenomenological anti-scientism that I want to defend could give rise to the following objections, which I hope to deal with elsewhere. Is the practice of science theoretical, that is, is it committed to a *vorhanden* view of things? Heidegger would seem to have to assent to this, although I remain less convinced. For example, is not a biologist involved with an experiment involved in the same way as Dasein is involved in being-in-the-world? I don't see why not, and therefore there is a question for me as to how one might understand science as a practice that could well be *zuhanden*, until the laboratory equipment breaks down of course. In this case, science would also be a praxis and not a theoreticism.

  • Doesn't my defence of Heidegger end up endorsing a primacy claim, namely that the *zuhanden* relation to the world *founds* the *vorhanden*? The question is, do I need this foundational, primacy claim? Couldn't I just make things easier for myself by arguing for a division of labour between the humanities and the sciences? No, because I think Heidegger is correct, namely that I think we do see the world first and foremost as *zuhanden* and the *vorhanden* is grafted on to that; "knowing is a founded mode of being-in-the-world".

  • Can or should the theoretical *vorhanden* attitude be extirpated from human activity? Not at all. Not all experiences of the *vorhanden* can be thought of as falling away from our everyday practices. There are some very practical experiences of the *vorhanden*, or some experiences of the *vorhanden* that are embedded in our practices. For example, to tell a joke is to ask the listener to look at things from a reflective, contemplative point of view. Many of our aesthetic experiences also have this form. They are theoretical experience. A life of pure praxis would be both unsustainable, reactionary, and dull.

- Finally, doesn't such an interpretation of Heidegger end up dividing the earlier work from the later work insofar as the picture of the relation of phenomenology to science seems to give way to quite another picture of science as defining the modern conception of the world in terms of the *Gestell*, where scientific explanation is reduced to the lawful experience of nature in the modern world, and is usually thought of in terms of calculation? That is, in his later work, it is difficult to see Heidegger's position as complementing a scientific conception of the world, but rather replacing it. If this is right, then this might begin to explain some of my doubts about the later Heidegger.
- The issue, to reiterate, is not science, but scientism; it is the latter and not the former that is pernicious.

2

HEIDEGGER'S *BEING AND TIME*

Reiner Schürmann

Introduction: situating *Being and Time*

The common thesis: Being and Time *and the philosophy of subjectivity*

Since its appearance in 1927, it has remained unclear what kind of book *Being and Time* is. Between the wars there were at least two prominent misreadings. The first we might call the "existentialist misunderstanding". *Being and Time* was read as if Heidegger wanted to express certain moods of absurdity predominant after the catastrophes that followed World War I. This is the reading from which Sartre took his lead, selecting some themes from *Being and Time*—Being-towards-death, dread, etc.—and developing them into a so-called "ontology of human existence".[1] The second misreading could be labeled the "anthropological misunderstanding" of *Being and Time*. Husserl wrote on the first page of his copy of *Being and Time: "Ist das nicht Anthropologie?"*. The names under which this type of reading long survived are Otto Friedrich Bollnow and Erich Rothacker. Certainly one must concede that *Being and Time* lies at the origin of what was called existentialism and that it significantly modified philosophical anthropology; yet, strictly speaking, it belongs in neither of these two categories.

Since the 1950s, a new thesis has emerged.[2] What Heidegger aimed at overcoming in *Being and Time* was the traditional understanding of man as one entity, one *res,* among others—endowed, not with chlorophyll as some plants, nor with wings or fins as some animals, but with "animal rationale". Man is that living being that possesses reason (or speech, since this is the Latin version of Aristotle's ζῷον λόγον ἔχον). Quite correctly, *Being and Time* was and continues to be seen as an attack against the uncritical division of things into those that are merely physical and those that also have a mind, into extended things and thinking things. This view, also correctly, emphasizes the concept of Dasein to show that it is polemically oriented against the picture whereby what is proper to man is wholly constituted by man's "specific difference" within a greater genus.

The common view, then, locates *Being and Time* within the tradition of the philosophy of subjectivity. On this reading, *Being and Time* would renew our understanding of the human subject—not through "existentialist" descriptions, nor through "anthropological" findings, but through a re-articulation of the relation between man and the world. To state it simply: Dasein means that man cannot be understood without his world, and correlatively that the world is always man's world. This signals the end of a solipsistic subject.

Let us investigate this common view a little further, since it is certainly not wrong, but insufficient. As you may notice, out of the 83 sections of *Being and Time,* 75 deal with an analysis of what Heidegger calls "Dasein", for which there seems to be no English equivalent. Eight sections, at the beginning, seem to constitute a somewhat broader introduction. Needless to say, after having labored through those very dense 75 sections, the first eight are often more or less forgotten. The common view states, still correctly, that Heidegger did not undertake the existential analytic for its own sake, but for the sake of what was sketched out in the first eight sections, the so-called "question of Being". Being is, so we are told, always man's Being—hence "*Da-sein*". Thus Heidegger renews the philosophy of subjectivity by exhibiting its ontological foundations, by means of "fundamental ontology".

What the common view cannot account for is that Being is not "always that of Dasein". Therefore, fundamental ontology does not simply show how Dasein grounds itself transcendentally. Or, stated otherwise, *Being and Time* does not merely clarify the "ontological" (and we would have to see what that word means in each context) meaning of intentional acts (which Husserl and Scheler already opposed to a philosophy of man that sees him as substance, thing, *ens creatum,* etc.) or the ontological foundation of the "person".

Thus, the common view holds that *Being and Time* "begins" with the de-substantialized subject. Indeed, *Being and Time* would only carry further this process of de-substantialization that had already begun with Kant and Hegel and was pushed further by Schelling and Kierkegaard. For all these authors the subject is actually no longer a *res,* a given thing.[3] Heidegger's point of departure is the notion of subject as "process" (*Vollzug*), and in this respect he could be said to belong to the post-idealist tradition. Such is his "place in the history of philosophy".[4] Roughly speaking, Hegel broke with the metaphysical tradition that views the subject as one being among other beings, instead sublating the subject in the infinite process of the spirit mediating itself and all things. Schelling, from this perspective, discovered the finitude of this spirit, its facticity, its "thrownness" even, since he saw finite spirit as being "thrown" (this is Heidegger's term) into existence, i.e. as not the master of its Being. Likewise, Kierkegaard described this facticity in terms of the imminence of death, dread, etc., but still in relation to an absolute. Heidegger, then, incorporates the facticity into the essence of man himself; Dasein is thrown into the world, but there is no thrower. In its process (*Vollzug*), the subject, considered in itself, is now utterly finite. This, as we shall see, is the meaning of "wholeness" or "totality" (*Ganzheit*). *Ganzheit* is not the sum total of traits belonging to Dasein, but its finite autonomy; its utter facticity, with no recourse to an infinite subject. Thus the title of

Heidegger's book, *Being and Time,* becomes clear. The meaning of the subject's Being is time; the subject's Being cannot be referred back to anything other than Dasein, out of which it would then "enter" into time. All these quasi-theological constructions are indeed eradicated by *Being and Time.*

The conclusion of the "common view" is that *Being and Time* can indeed be viewed as a work of the philosophy of the subject.[5] It is the final work of such a philosophy, since in *Being and Time* the subject becomes radically autonomous, accomplishing itself. In Heidegger's terminology, thrownness (*Geworfenheit*) and project (*Entwurf*) are structures of Dasein. *Being and Time,* in other words, is the attempt to understand the subject completely out of itself, neither in comparison with other things nor in relation to some supreme subject.

All this is correct, but this thesis does not operate with the understanding of the word "Being" that Heidegger explicitly works out. "Being" is *not* primarily man's (Dasein's) Being. To balance the common thesis, we have to take a brief look at where *Being and Time* really begins. It is important to understand clearly that in *Being and Time* Heidegger is preoccupied with the question of Being as such— whatever that will turn out to mean—and only *therefore* with the question of Dasein.[6] This is important, among other things, for the unity of Heidegger's writings. William Richardson introduced, in 1963, the distinction between Heidegger-I and Heidegger-II: the young Heidegger, from this interpretation, was preoccupied with questions of existence, of the subject, and the latter with that of Being *qua* Being.[7] Although Heidegger does speak of a "turn" or reversal (*Kehre*) in his writings, he explicitly denies a break in his thought and says that he never abandoned the intent of *Being and Time.*[8] It is true, though, that when he speaks of *Being and Time* in later publications he always refers to the first eight sections. Thus, these first 40 pages are proof that Heidegger's work is a unity, that there are not "two Heideggers". This also makes it quite clear that, more or less explicitly, we will have to take into account the later writings when interpreting *Being and Time.*

Being and Time *as retrieval*

The first thing one notices when opening *Being and Time,* is that it takes up a philosophical issue not from Hegel, Schelling, Kierkegaard, or Husserl (aside from the dedication to Husserl), but from Plato, whose *Sophist* provides the epigraph for the book. The first paragraph of *Being and Time* speaks of the "question of Being" and says that it "sustained the avid research of Plato and Aristotle" (SZ 2, JS 1). The first section is entitled "The Necessity of an Explicit Retrieval [*Wiederholung*] of the Question of Being" (ibid.).[9] Rather than continuing the tradition of German Idealism and of transcendental phenomenology, Heidegger's book wants to retrieve something that, according to him, "ceased to be heard *as a thematic question of actual investigation*" (ibid.) from Plato and Aristotle onward. From the outset, Heidegger situates *Being and Time* in continuity with Greek Antiquity rather than with modernity and contemporary philosophy. *Being and Time* is thus altogether a retrieval, albeit in a complex way, as I will explicate in the following paragraphs.

What is to be retrieved, according to the epigraph, is "the question of the meaning of Being". How so? In the *Sophist,* a Stranger is speaking with Theatetus. The Stranger addresses a group of men. These are described as the men who try to understand how many and of what nature "beings" are. They proceed to answer their own question: namely through "storytelling". They speak of warfare and love, and say that in such happenings beings come to be; that they arise from other beings and pass away into other beings. In other words, people[10] grasp beings through their generation and corruption. Heidegger remarks a little later:

> The first philosophical step in understanding the problem of Being consists in avoiding the μῦθον τινα διηγεῖσθαι [*Sophist* 242c], in not "telling a story", that is, not determining beings as beings by tracing them back in their origins to another being—as if Being had the character of a possible being.
>
> (SZ 6, JS 5)[11]

In the *Sophist,* the Stranger admits his perplexity with regard to the question of Being, and then begins another dialogue with Parmenides. Here a true discussion of Being occurs. So, what is retrieved at the beginning of *Being and Time* is the transition from a storytelling to Parmenides' thought of Being. More simply: it is the thought of Being as addressed by Parmenides, the Stranger, and Plato that is retrieved.

But then, it is something in ourselves that apparently has to be retrieved. To cite the same epigraph: "...[a]re we today even perplexed at our inability to understand the expression 'Being'? Not at all" (SZ 1, JS xix). Thus, the basic θαυμάζειν of Platonic thinking is to be retrieved, or Leibniz's question (quoted at the beginning of the *Introduction to Metaphysics*): "Why are there beings rather than nothing?" (EM 1/1). Our capacity for wonder has to be retrieved. The beginning of *Being and Time* lies neither in modern nor ancient philosophy, but in us. And we shall see that the prerequisite of a certain disposition in he who philosophizes—of a practical a priori, so to speak—remains at work not only throughout *Being and Time,* but also throughout all of Heidegger's writings. When Heidegger here speaks of the *Sinn* of Being, one may perhaps also hear: "a 'sense' for the question of Being, a sensibility, has to be developed in us" (SZ 1).[12]

But retrieval is still more complex. In Section 1 we read about three prejudices: "We wish to discuss these prejudices only to the extent that the necessity of a retrieval of the question of the meaning of Being becomes clear" (SZ 3, JS 2). These three prejudices are:

- Being is the most general of all concepts
- it is not definable
- it goes without saying.

Heidegger then writes:

> [a]n enigma lies a priori in every relation and Being toward beings as beings. The fact that we live already in an understanding of Being and that the meaning of Being is at the same time shrouded in darkness proves the fundamental necessity of retrieving the question of the meaning of "Being".
>
> (SZ 4, JS 3)

What is to be retrieved here? An a priori enigma. This is strange, since we had just been told that, since Plato, our philosophers have forgotten the question of Being. If there now lies, a priori, an enigma in every comportment with regard to beings, what happens to forgetting and remembering? First, there was a kind of accusation directed at us: you are not even embarrassed that you don't understand the expression "Being". But now, the very lack of embarrassment about not understanding is called an a priori enigma. There is not much that we can do about an a priori enigma; this seems at least to be implied by the expression "a priori". If "we" have forgotten the question of Being, then Heidegger speaks of an empirical, "ontic" state of affairs. But if there is an a priori enigma in our relation to beings, he speaks of what will have to be called an "ontological" state of affairs.[13] In this light, one may have to translate *Sinn des Seins* not as "sense" of the question of Being, but properly as "meaning of Being". This meaning, necessarily shrouded in darkness, seems to be what has to be retrieved—but the complications do not end here.

In Section 2, yet another retrieval is operative. That section again has three parts: Heidegger distinguishes between "what is asked" (*das Gefragte*), "what is interrogated" (*das Befragte*), and "what is to be ascertained" (*das Erfragte*). What is asked, or asked about, is Being itself; what is interrogated is human Dasein; and what is to be ascertained is the "Sinn" of Being. What is asked about is called by Heidegger: "Being, that which determines beings as beings" (SZ 6, JS 4). Being determines. He also speaks of the "essential determination by Being itself" (SZ 6). It seems that this determination *by* Being is what has to be retrieved. In his last lecture, in 1962, Heidegger said: "We want to say something about the attempt to think Being without regard to its being grounded in terms of beings" (SD 2/2). When compared with the beginning of *Being and Time,* this says nothing new: Being, as determining beings, is not grounded in them. In this last approach, then, Being itself seems to initiate *Being and Time.* This impression is enforced here by the first allusion to "transmitted theories and opinions about Being" (SZ 6, MR 25). Here, it is suggested that it is not a matter of chance that between Plato and ourselves the question of Being has been "forgotten", that the responsibility of that forgetfulness lies, so to speak, with Being itself. That this line of reading does not lead to any "mythologizing" of Being cannot be shown now. But the impression that Being itself begins *Being and Time* is reinforced by the third meaning of "*Sinn*", namely "directionality". We shall see that, in this way, time is introduced into the question of Being.

Table 2.1 Three ways of understanding the retrieval and "*Sinn*"

The beginning to be retrieved	Understanding of "sense"
• "Tahumazein" in Plato, in us • A priori enigma • Being	• Sensibility for the question • Meaning • Directionality

We can take the first two of these four facets of the retrieval together. These two facets do indeed speak of the same transition to serious questioning, first in Plato, then in us. We thus have three ways of understanding the retrieval, and correspondingly three ways of understanding "*Sinn*" (Table 2.1).

For the sake of completeness it must be added that the concept of retrieval (*Wiederholung*) has two more meanings in *Being and Time;* one related to the structure of the book, and the other to its very content, the understanding of the Being of Dasein as Care (*Sorge*).

According to the outline in Section 8 of *Being and Time,* the work was to consist of two main parts, each of which was to contain three divisions. Only the first two divisions of the first part are contained in what we now have under the title of *Being and Time.* The third division was given as a lecture course and later published under the title *Grundprobleme der Phänomenologie.*[14] The content of Part Two, to be entitled "Basic Features of a Phenomenological Destruction of the History of Ontology on the Guideline of the Problem of Temporality" (SZ 39, JS 35), has been carried out in the many books published after *Being and Time*—although not exactly in the way announced in *Being and Time.* Division Two of *Being and Time* "retrieves", we are told, the content of Division One. Its task is "to retrieve (or repeat) our analysis of Dasein in the sense of interpreting its essential structures with regard to their temporality" (SZ 304, MR 352). Thus, the relation between Division One (the Analytic of Dasein, strictly speaking) and Division Two (entitled "Dasein and Temporality") is a relation of retrieval from the point of view of time.

The last meaning of the term retrieval has to do with what will be called "authentic resoluteness".[15] *Being and Time* is not the mere offspring of the philosophy of subjectivity[16] because it is altogether a retrieval—that is, it makes explicit and thematic something that has been operative throughout the whole of the history of philosophy, something that is also operative throughout man's life wherever it occurs. To "retrieve" a problem is nothing other than making it explicit and thematic. One retrieves what has been there "always already" (*immer schon*). The place of *Being and Time* in the history of philosophy can only be understood through this project of retrieval. It aims to articulate what has remained essentially unarticulated in all previous Western philosophy: not a small affair—"the most difficult thought of philosophy" is to "think Being *as* Time" (NI 28/I 20).

The aporia in Being and Time

In the "Letter on 'Humanism'", Heidegger writes: "It is generally supposed that the attempt in *Being and Time* ended in a blind alley. Let us not comment any further on that opinion" (BH 173/222/261). Seen from the point of view of the problematic of Being as quoted a little earlier, "We want to say something about the attempt to think Being without regard to its being grounded in terms of beings" (SD 2/2), the very table of contents of *Being and Time* spells out an aporia. The subject matter to be investigated is "Being itself", "without regard to its being grounded in terms of beings". This subject matter is stated clearly in the introduction to *Being and Time.* And in the "Letter on 'Humanism'" Heidegger continues: "The thinking that hazards a few steps in *Being and Time* has not advanced beyond that publication even today" (BH 173–4/222/261). On the other hand, in 75 out of 83 sections, Being is investigated only as the Being of Dasein.

This aporia is stated clearly in Section 2: "To work out the question of Being means to make one being—he who questions—perspicuous in his Being" (SZ 7, JS 6). The basic problem that one encounters in trying to understand *Being and Time* is the following: in order to work out Time as the meaning or directionality (*Sinn*) of Being, Heidegger ends up working out temporality as the meaning or directionality of Dasein. This is the reason why *Being and Time* had to remain truncated. It is not by chance that (aside from the lectures *Grundprobleme der Phänomenologie*) the title "Time and Being", which was to cover Division Three of Part One (SZ 39), is also the title of Heidegger's penultimate publication. This lecture, *On Time and Being,* was given in 1962. It shows the kind of unity that exists in Heidegger's writings: they constitute a whole, but not a systematic whole. Rather from *Being and Time* to *On Time and Being,* the unity is one of itinerary. The lecture *On Time and Being* is *not* what was planned as Division Three; however, it does treat the topic of the third division appropriately, but only after three and a half decades of what Heidegger calls a "path of thinking". I will quote at length from a remark Heidegger made about the lecture *On Time and Being* and its relation to Division Three of *Being and Time:*

> In the structure of the treatise *Being and Time* (1927), the title "Time and Being" characterizes the third division of the first part of the treatise. The author was at that time not capable of a sufficient development of the theme designated in the title "Time and Being". The publication of *Being and Time* was interrupted at that point.
>
> What this text contains, written three and a half decades later, can no longer be a continuation of the text of *Being and Time.* The leading question has indeed remained the same, but this simply means: the question has become still more questionable and still more alien to the spirit of the time.
>
> (SD 91/83)

It will ultimately become clear, I hope, that this aporia is not a matter of Heidegger's philosophical capability. In fact, there are strong reasons for preserving the aporia. Furthermore, Heidegger's decision not to publish any subsequent books (after *Being and Time* he only published collections of lectures and essays) has very little to do with the personality of Martin Heidegger. His work, and first of all *Being and Time,* stand historically on a threshold where systematic philosophy is no longer, or not yet, possible. The fragmentary character of his writings indicates what he calls a historical "determination by Being" (SZ 6).

Thus we have to say that *Being and Time* is not a systematic work, and that the entirety of Heidegger's corpus does not constitute a system. His works are unified solely by a path of thinking, a path that, in a sense, leads nowhere. In fact, he later published a book under the title *Holzwege,* woodpaths. This work's epigraph states:

> Wood is an old name for forest. In the wood are paths which mostly wind along until they end quite suddenly in an impenetrable thicket. They are called "woodpaths". Each goes its peculiar way, but in the same forest. Often it seems as though one were like another. Yet it only seems so. Wood cutters and forest rangers are familiar with these paths. They know what it means to be on a woodpath.[17]

In her commemorative article "Martin Heidegger at Eighty", Hannah Arendt wrote:

> The metaphor of "wood-paths" hits upon something essential—not, as one may at first think, that someone has gotten onto a dead-end trail, but rather that someone, like the woodcutter whose occupation lies in the woods, treads paths that he has himself beaten; and clearing the path belongs no less to his line of work than felling trees.[18]

Now let us once again reflect on the way *Being and Time* begins. It begins with a quotation from the *Sophist* and discussion about the forgottenness of the question of Being since Plato and Aristotle. Instead of forgottenness we might have to say something like "trivializing". The problematic of Being has been "co-opted", so to speak, by interests—the interest in a supreme being (Middle Ages), the interest in the regulation of behavior ("natural law" as derived from an ontology), and the interest in scientific knowledge (since Kant). This is what is meant by forgottenness. In these uses, the question of Being has become subordinate to other questions. Re-enacting the Stranger's perplexity would now mean being astonished by the itinerary that the question of Being has undergone since it was first raised in Greece.

Let us push this a little further. The metaphor of "wood-paths" hits upon something essential, not only in Heidegger's own itinerary of thinking. The "itinerary of Being"—Heidegger is particularly fond of phrases like this, but they are not without traps—would itself be something like a woodpath. What does this mean?

First of all, it means that the span from the initial question to its retrieval does not lead to some kind of *telos*. In the language that we have already encountered, the "determination" of beings by Being is not teleological. There is no apotheosis that will arise from the retrieval. We must say even more: the concealment of the question of Being has its root in the way that this question was first raised by the Greeks. Moreover, we shall eventually have to understand that concealment belongs essentially to Being, at least as it is thought in *Being and Time*.

That the itinerary of the question of Being resembles woodpaths also means that the entire path of philosophy since the Greeks is an errancy—not an error, but a wandering. This, too, has tremendous consequences, which I only gesture toward here: "Who thinks greatly must err greatly."[19] The aporia, as we have stated it, is such a necessary errancy since to think Being, purely and simply, is impossible. Thus, there is a good reason for Heidegger to write: "Let us not comment any further on that opinion", after having reported that "it is generally supposed that the attempt in *Being and Time* ended in a blind alley". The good reason is, to say it again, that concealment is essential to Being, that the path of Western thought, the forgottenness of the question of Being, is one instantiation of that essential concealment; and thus that Heidegger's own path of thought is another instantiation of it. (Heidegger himself would go so far as to say that his political error in the 1930s is an instantiation of Being's concealment.) Such is the answer to the aporia stated earlier—the answer as it can be gathered from the metaphor of "woodpaths" or, more technically, from what we shall have to call "aletheia".

We set out trying to "situate" *Being and Time* within the history of philosophy. It appears that its site is precisely the entire history of philosophy, since its purpose is to gather up the problematic that "makes" that history. It encompasses that history quite as Plato encompasses that history (if it is at all true that Western philosophy is "but a series of footnotes to Plato", as Whitehead says). *Being and Time,* from the start, situates itself as the counterpart to Plato and Aristotle; that is its "site" in the history of philosophy.

Dasein as the exemplary being for the retrieval

To retrieve the problematic of Being, Heidegger, at the beginning of *Being and Time,* proceeds in a twofold way, comprised of both a negative and positive aspect. The negative aspect consists in discarding—dismantling, rather—received opinions that cover up the question of Being. The positive aspect consists in discovering one being in terms of which the question of Being may appropriately be raised. As we have already seen, that particular being will be ours, i.e. Dasein: "To work out the question of Being means to make one being—he who questions—perspicuous in his Being" (SZ 7, JS 6). But to understand why and how Dasein can be the exemplary being for the retrieval, we first have to see why and how received opinions provide no means for a retrieval. The reason for this is that received ontologies function as foundations for knowledge. The first question to be raised will therefore be: in what sense is the retrieval "fundamental"?

The idea of 'fundamental' ontology in Being and Time

'Prejudices' about 'Being'

At the end of the very first paragraph of *Being and Time* the situation is as follows. The question of Being, first raised by Plato and Aristotle, has fallen into forgottenness.[20] It has, however, remained active in its very concealment, under "various distorted and 'camouflaged' forms" (SZ 2, JS 1)—it led a hidden life, so to speak. The paragraph concludes with the following words: "[w]hat then was [with Plato and Aristotle] wrested from the phenomena by the highest exertion of thinking, albeit in fragments and first beginnings, has long since been trivialized" (SZ 2, JS 1).

The methodological hint is clear: to overcome the forgottenness, we must, negatively, labor against this trivialization and, positively, do what Plato and Aristotle did, i.e. wrest the question of Being from the phenomena—that is, from one exemplary phenomenon.[21] This struggle against trivialization may not entirely spare Plato and Aristotle. Indeed, as the above lines state it, they raised the question of Being only "in fragments and first beginnings"; and the next paragraph adds, "[o]n the foundation of the Greek point of departure for the interpretation of Being, a dogma has taken shape" (SZ 2, JS 1). We might add: the dogma is that Being can yield a "science", a "first philosophy".[22] Such a dogma of a "grounding science" stands in opposition to the task of wresting the question of Being from phenomena—in other words, phenomenology is opposed to an ontology that "founds".

Heidegger, in order to free the space for the retrieval, suggests that this dogma takes three forms, adding that these prejudices "have their root in ancient ontology itself" (SZ 2–3, JS 2).

1   The first form of the dogma is the claim that 'Being' is the most 'universal' concept. According to this prejudice, the mechanics of definition (by genus and specific difference) would be able to teach us something about Being. All genera have a limited extension; that is why they are many. But Being would be the genus of all genera. Heidegger quotes Aristotle's *Metaphysics* to discard this prejudice: the universality of Being is not that of a genus. Aristotle's reasoning is well known: a specific difference such as "to think", "to speak", "to laugh" is always added to a genus from the outside so as to differentiate it, e.g. to the genus "animal" or "living being". But if Being were the highest genus, one would have to go outside the concept of Being to differentiate it—which is contrary to the premise. As Heidegger says, "[t]he 'universality' of Being 'surpasses' [transcends] the universality of genus" (SZ 3, JS 2). One must read this paragraph on Aristotle carefully, as well as the two passages from the *Metaphysics*. The paragraph ends: "To be sure, Aristotle did not clarify the obscurity of these categorial connections" (SZ 3, JS 2). It is in the *pros hen* relation that Aristotle at the same time wrests the question of Being from phenomena and obscures it dogmatically. Indeed, this *pros hen* relation

is then taken up by Aristotle to become the chief tool for analyzing entities in the *Physics, Ethics, Politics* and in the *Organon.*

2   The second form of the dogma is that "the concept of 'Being' is indefinable" (SZ 4, JS 2). As with the first prejudice, Heidegger concurs with this view as long as it is understood properly: Being is universal, even transcendent or transcendental, but not a genus. Likewise, Heidegger agrees that Being is indefinable, since "'Being' cannot be understood as one among the beings" (ibid.). But not because the method of defining by genus and specific difference would capitulate before Being's generic extension. The reason why Being is indefinable has nothing to do with the success or failure of the definition by genus and specific difference, but rather with the way we have to think of determinateness.[23] Being is to be determined, but not in the way a definite being can be determined. "But does it follow from this that 'Being' can no longer constitute a problem? Not at all" (SZ 4, JS 3). Thus, what has to be retained from this second prejudice makes Being all the more questionable.

2   The third form of the dogma is that "'Being' is the self-evident concept" (ibid.). Here again, this claim must be properly understood. We constantly use the word 'being' and the verb 'to be'—as Aristotle observed. And we remember that "in any relation to beings there lies a priori an enigma" (SZ 4). So we move and speak always within an understanding of Being, a pre-understanding, to be precise. But this pre-understanding is ours "*zunächst und zumeist*" (proximally and for the most part) (SZ 16, MR 37). Later this will be called a result of Dasein's "falling" (*Verfallen*) into average everydayness. So, this prejudice has to be rejected if it is meant to imply that there is no need for an inquiry into the question of Being, that Being is not a question.

The three prejudices teach us something positive, namely, that we are "always already" (*immer schon*) acquainted with what Being means, but at the same time that this meaning remains an enigma. The three prejudices have to be discarded if they are meant to imply that we already know enough about Being by this pre-reflexive acquaintance. Thus, Heidegger discriminates, in each of these prejudices, between a sense that is genuinely phenomenological and one that obscures the "phenomenon" of Being. But the problem remains: it is the "foundational" character of Being, of the experience of Being, that "obscures".

Pre-understanding and the point of departure of the retrieval

For Plato, the question of Being emerged in the transition from apparent knowledge of the meaning of Being to perplexity with regard to its meaning. Aristotle raised the question of Being by observing that the copula is used all the time and in many forms whenever we speak, but that these many ways of saying "*on*" all refer toward the one, *pros hen,* which is that of substance. The common prejudices, finally, say that we all know what Being means. One is thus left with the impression that we ourselves are somehow implicated in the question. In the move to perplexity, *we* are involved; in the manifold usage of the copula, *we* speak; and in the opinion that everything is clear about Being, *we* surmise. Perhaps the problem of "foundation" leads to "us".

This state of affairs reveals two important points, namely, that all questioning is guided in advance by what is sought, and—most importantly—that we shall have to turn to ourselves in order to learn more about the question of Being.

We do not insist that to ask the question of Being one must focus solely on these three structural moments of the reciprocity between seeking and what is sought, namely: that which is asked about, i.e. Being; that which is questioned, i.e. the particular being that we are; and that which is to be ascertained, i.e. the meaning of Being. The issue is rather the following. To raise the question, to raise any question, is to look for some determinateness of that with which we are already familiar. This sounds like common sense, yet it includes the three elements just mentioned: asking about something, looking for a particular point of attack, and then "determining".[24] Heidegger says of conceptual determination that "[w]e are always already involved in an understanding of Being...[But] we do not *know* what 'Being' means"; i.e. we are not "able to determine conceptually what the 'is' means" (SZ 5, JS 4).[25]

One notices the opposition between knowing and understanding in this quote: "*This average and vague understanding of Being is a fact*", he says, but "we do not *know* what 'Being' means" (SZ 5, JS 4). Here, as in so many other instances, the influence of the hermeneutical tradition, and particularly Dilthey, shines through. The end of the book is a lengthy return to Dilthey and his friend Count Yorck von Wartenburg (the book does not really end, it merely falls apart into long quotations). But here one can also see that Heidegger is strangely anti-hermeneutical. The task of "determination" is to lead us from understanding (pre-understanding) to knowing. This is only one example of the many instances that caused Heidegger to say later of *Being and Time* that it had not, as yet, entirely disentangled itself from metaphysics.

To make the question of Being "questionable" again is to show that in the three "prejudices" there is something wanting, that the indeterminacy of our acquaintance requires further determinations. But to take that question up as a question also means that these further determinations will only be multiple "inroads", woodpaths, and hence, to quote Nietzsche "'the way'—that does not exist".[26] Heidegger indicates three "ways" that he has followed in order to make the question of Being questionable again: the first way is according to the "meaning" of Being (*Being and Time*); the second is according to the "truth", *aletheia,* of Being (historical constellations of truth); and the third is according to the "topology" of Being (again with regard to constellations, no longer of truth, but rather of "presence").[27]

Let us now gather the "positive" determinations of the starting point of *Being and Time:*

1   The pre-understanding that is "always already" operative in what we do and think, the "a priori enigma".

2   The pre-understanding as specifically "ours" today, i.e. at that moment in the
history of philosophy where "we deem it progressive to give our approval to
'metaphysics' again" (SZ 2, MR 21), but in such a way that the purported
subject matter of metaphysical ontologies has been trivialized to the extreme.
3   The pre-understanding that is guided neither by "substance" nor by the "subject", but by "average everydayness", by that which occurs in daily existence
"*zunächst und zumeist*", proximally and for the most part.

The point of departure for the retrieval is thus simultaneously a priori, historical,
and "existential". Since this point of departure is a priori, it requires a conceptuality inherited from transcendental philosophy, which utilizes peculiar kinds of
categories. Furthermore, since the point of departure is historical—located at the
counter-position to Plato and Aristotle and at the moment of completion of the
philosophy of subjectivity—it requires a conceptuality whereby both ousiology
and transcendental philosophy are appropriated and explicitly brought back to
their hidden roots in the problematic of Being. Finally, since the point of departure is in life, i.e. is "existential", these categories will be neither those of *ousia*
nor those of the transcendental subject, but those of everyday life. Because these
categories are, precisely, of existence they will have to be "*existentialia*".

If the point of departure is thus pre-understanding in this threefold mode, we
should perhaps replace the talk of "aporia" in *Being and Time* by that of a "circle".
Indeed, pre-understanding is a mode of our Being; the determination of that vague
acquaintance is "questioning"—a more determinate mode of our Being. We can
ask about Being only in making our own questioning questionable. The movement
of departure is thus questioning our questioning or making problematic our vague
acquaintance with Being. What is clear is that we cannot question Being as if it
were "out there", something other in which we are not also "involved", as if it were
"a being". But the circle of questioning is not a vicious one: "What is decisive is
not to get out of the circle, but to get into it in the right way. The circle of understanding is not an orbit in which any random kind of knowledge may move; it is the
expression of the existential *fore-structure* of Dasein itself" (SZ 153, MR 195).

The threefold modality of pre-understanding—a priori, historical, existential—exhibits the "*wesenhafte Betroffenheit*", the way we are concerned and
engaged by such an inquiry. "The way our questioning is essentially concerned by
what is questioned, belongs to the innermost meaning of the question of Being"
(SZ 8). The old matter of ontology is now seen as engaging *us*—that is a retrieval
of the *ti to on,* which is historically possible only after Kant and the movement of
examination of the subject that he inaugurated. In the circular structure of pre-
understanding "something like a priority of Dasein has announced itself" (SZ 8,
JS 7). Thus we can say that the point of departure for the retrieval is Dasein itself.

Heidegger introduces the word "Dasein" in the following way: "This being,
which we ourselves are and which has questioning as one of its possibilities of
Being, we denote as Dasein" (SZ 7).[28] Dasein is thus the point of departure of the
retrieval, because it is "essentially" concerned with the question of Being.

REINER SCHÜRMANN

In Section 2, Heidegger opposes a "bad" circularity in the structure of the question as it arises from pre-understanding. The bad circle claims that to raise the question of Being, i.e. to analyze Dasein, one must already know the answer to the question of Being. Heidegger's argument here is that there is, in fact, no such circle because we live always already in a vague acquaintance with the meaning of Being. This acquaintance suffices to direct our questioning about Dasein, so that instead of a structure of circling, we have one of "determining", i.e. determining the understanding that "belongs to the essential constitution of Dasein itself" (SZ 8, JS 6).

The twofold priority of the 'Question of Being'

We are now in a position that allows us to resolve the ambiguity in the talk of "foundational" ontology. Aristotle opens, and at the same time obscures, the question of Being by making it the object of a first science that, albeit "sought for", is nonetheless known to be "first"—and "first" here means that this particular science founds or grounds all others.[29] Against this discourse about Being, Heidegger opposes another, which he calls "fundamental ontology". We understand from what has been said to this point that fundamental ontology will not found, ground, justify, legitimize, or stabilize any kind of knowledge; it cannot be used for securing criteria for doctrines about human behavior, private or public, for logic, or for science in the modern sense. So, in what sense does questioning about Being lead to a "fundamental ontology"? What is the meaning of this "fundamental"?

Section 3 is entitled "The Ontological Priority of the Question of Being". The text abounds with phrases like *"Grundbegriffe"*, *"Grundlagen"*, and *"Grundverfassung"*. It is clear from the start that the question of Being is going to operate as a "grounding". But what kind of grounding? Is it a grounding as in metaphysics, i.e. the securing of an ultimate, indubitable, enduring, normative First?

Let us look at the text. Heidegger speaks of "regions", *Bezirke, Felder*— regions of scientific investigation. These are, for instance, "history, nature, space, life, human being, language, and so on" (SZ 9, JS 7). Each of these domains needs to be "grounded", that is, their realm, the totality that they constitute, has somehow to be justified. It is an old discovery that mathematics cannot ground mathematics, that physics cannot draw the line that circles its domain of validity, and so on. Philosophers have long claimed that it is their task to provide the foundations that make sciences possible. Thus, ontology is more primitive, primordial, and authoritative than the sciences.[30] "Ontological inquiry is more originary than the ontic inquiry of the positive sciences" (SZ 11, JS 9). "Originary" indeed: the possibility of demarcating regions for positive investigations is what ontology provides. For instance, Heidegger says that Kant's "transcendental logic is an a priori 'logic of things' [*Sachlogik*] of the domain of Being called nature" (SZ 11). That is to say, the *Critique of Pure*

69

Reason, in this view, anchors and legitimizes the sciences of nature, and primarily physics (we shall soon examine an interpretation of the *Critique of Pure Reason* that is diametrically opposed to this).

Ontologies thus constitute a step beyond the empirical, allowing for regional grounding. Ontologies provide the "a priori condition...of the sciences which investigate beings of such and such a type" (SZ 11, JS 9). Such ontological inquiries—as performed by Aristotle with regard to the *Organon,* by the Medievals with regard to a *Summa,* and by Kant with regard to the metaphysics of nature—are "investigations into the Being of beings" (ibid.). Indeed, it says that beings such as heavy bodies are "physical", beings such as roaches "zoological", beings such as phonemes "linguistic", etc. Ontologies provide "a genealogy of the different possible ways of Being, construed deductively" (SZ 11, JS 9). In short, ontologies ground sciences by assigning them their regions of inquiry. In that sense they are foundational, e.g. they demarcate "nature" for the possibility of natural sciences.

Heidegger then writes, "[i]t is true that ontological inquiry is more originary than the ontic inquiry of the positive sciences. But it remains naive and opaque if its investigations into the Being of beings leave the meaning of Being in general undiscussed" (ibid.). What can this mean? Assuredly, a second transcendental step backward, not only from the positive sciences to their ontological conditions, but again "to the condition of the possibility of the ontologies which precede the ontic sciences and found them" (SZ 11, JS 9). This is surprising. It seems that we are invited to step back to some yet more solid rock, some yet more grounding ground.[31]

Let us focus on this 'double step backwards' from the positivity of the empirical sciences. In his book *Kant and the Problem of Metaphysics,* Heidegger uses a slightly different language to describe the same two-step move (and here we have the other interpretation of the *Critique of Pure Reason,* in which "critique" does not mean "*Sachlogik*", but the step back from the regional ontology of natural things toward the origin of such an ontology—the two approaches to Kant are irreconcilable). Table 2.2 compares the 'double step backwards' from the positivity of the empirical sciences in both *Being and Time* and Heidegger's book *Kant and the Problem of Metaphysics.*

In *Kant and the Problem of Metaphysics* we read (in the very first section): "The task of the following investigation is to explicate Kant's *Critique of Pure Reason* as a laying of the foundation of metaphysics in order thus to present the problem of metaphysics as the problem of fundamental ontology" (KPM 1/3/1).

Table 2.2 The 'double step backwards' in KPM and SZ

Kant	KPM	SZ
• Physics	• Empirical sciences	• Ontic knowledge
• Metaphysics of nature	• Metaphysics	• Ontological knowledge
• Critique of pure reason	• Laying the foundation	• Fundamental ontology

Laying the foundation of metaphysics does not mean "putting a foundation [*ein Fundament*] under this natural metaphysics or replacing one already laid by a new one" (KPM 1/4/1). Rather, fundamental ontology or laying the foundation of metaphysics means:

> ... the projection of the building-plan itself ... The tracing of the architec-
> tonic limits and design of the intrinsic possibility of metaphysics ...
> Bringing to light the originative origin [*die Ursprünglichkeit des Ursprungs*] of metaphysics ... This basic originality can be essentially
> understood only if it is brought into the concrete happening of the act of
> origination [*Geschehen des Entspringenlassens*], that is, if the laying of
> the foundation of metaphysics is retrieved.
>
> (KPM 2/4–5/2)

From this parallel text it should be clear that the retrieval of the laying of the foundation is an act of gathering up a process: a happening whereby the realm is first opened up in which metaphysics may then grow and sciences be justified. A happening is to be retrieved. What happening? "The concrete happening of the act of origination". What does that mean? In the light of the link between the question of Being and Dasein, "origination" will probably somehow refer to that link. The happening to be retrieved is the very birth of that link, it is the "*Sinn*" that Being has for Dasein, the constitution of that "*Sinn*". In a language to be clarified later, what is to be retrieved is the event by which Dasein opens a world, i.e. its "Being-in-the-world".

I only want to insist on the event-character at this point. This is the trait that shows most clearly the difference between metaphysical ontologies (which found sciences) and phenomenological ontology (which is said to be the foundation of metaphysical ontologies). The step from metaphysical ontologies to fundamental ontology is of a different kind from the step from the sciences to metaphysical ontologies. Everything hinges on this heterogeneity in the 'step backwards'.

Indeed, the very way in which *Being and Time* can be said to be a work of tran-scendental philosophy hinges on this heterogeneity. It is altogether an inquiry into the conditions of possibility of experience—but stepping from one type of ground to another type of "ground". It should now be clearer why it is not sufficient to read *Being and Time* simply as the latest offspring of the philosophy of the transcenden-tal subject. Its transcendentalism leads back behind the subject as origin of the forms that justify science. It leads to a reciprocity between man and not-man; to the breakdown of the isolated subject; to Being-in-the-world as originative of the transcendental subject that in turn is originative of science. This event of origina-tion—of "leaping ahead [*vorspringen*]", of "disclosing [*erschließen*]" (SZ 10, JS 9)—is Heidegger's transcendentalism in *Being and Time*.

The next section gives the key formulation of the link between the question of Being and Dasein. It also provides the second type of priority of the "question of Being" over any other possible questioning or investigation.

"Dasein is a being that does not simply occur among other beings. Rather it is ontically distinguished by the fact that in its Being this being is at stake *in* its very Being" (SZ 12, JS 10). The redundancies are only apparent. And they are less redoubtable than they sound. The phrase "Dasein is that among beings which, in its Being, is concerned *about* its very Being [*diesem Seiendem in seinem Sein* um *dieses Sein selbst geht*]" (SZ 12) reappears like a refrain throughout *Being and Time*. This simply means that Dasein is such that its Being is at issue, that it always and everywhere comports itself to its Being *as* something at issue or at stake. Such is the link between Dasein and the question of Being. Dasein is Dasein only insofar as its Being is an issue for it. This does not mean that Dasein is Dasein only if it explicitly raises the question of Being, for regardless, its Being is always at issue. The questionableness of Dasein's Being is continually lived through.

This link is so intimate that "Dasein" receives its very name from it.[32] The word "Dasein" stands for the program (with the difficulties that accompany it and which I have indicated) of retrieving the question of Being by questioning the Being of ourselves. Dasein has *a priority* among beings; this priority is the indissoluble link between interrogating ourselves (*befragen*) in order to put Being into question (*fragen, das Gefragte*) so that Being's meaning can be ascertained (*erfragt*). The link expresses itself in the term "meaning". By "meaning of Being" we can only understand "meaning of Being for us". It is this reference to ourselves in the vocabulary of meaning that made Heidegger later renounce such vocabulary. Indeed, the talk of "*Sinn*" was prominent among the neo-Kantians who were his teachers, mainly Rickert, but also Cassirer and, of course, Dilthey.

The second priority of the question of Being lies in its meaning for us; that is its ontic priority. But why use Dasein to refer to "us"? It goes without saying that Dasein is not something other than "man", but to say "man" instead of Dasein would be to operate with a pre-understanding that precisely does not involve the problematic of Being. The pre-understanding carried by the word "man" is tied to the obfuscating element in classical metaphysics,[33] to that element due to which, since Aristotle, the question of Being has remained trivialized. We recall that this trivialization operated, among other things, in the mechanics of the definition by genus and specific difference. The pre-understanding that the word "man" brings with it is such an obfuscating instantiation of the definition. To say "man" is to say "reason", *animal rationale,* crown of creation. Or, the pre-understanding that accompanies "man" is some version of the formal "I", the *cogito,* or the transcendental apperception. Here again we lose the primordial character of everydayness. Thus, Dasein stands for the link between the question of Being and the question of *our* Being, as it is tacitly operative in everyday life.

"Rational animal" thematizes man as a certain type of substance differentiated by reason; "Ego" thematizes man as a certain unity of functions and thus no longer as a substance, a thing; "Dasein" thematizes man as involved, necessarily and always, in the meaning of Being since for Dasein its own Being is always an issue. Dasein means the "there" of "Being-there-in-the-world". This word already suggests the angle from which Heidegger approaches man in *Being and Time,* namely that of our embeddedness in a finite world in which our Being is always at issue. The ontic privilege of the question of Being is thus that it has its site in one kind of being, ourselves, and not in others. Other kinds of beings have an environment or a milieu, but not a world, because for them their Being is never at stake.

What we called the link between Dasein and the question of Being is called by Heidegger a relation. Dasein is such "that in its Being it has a relationship of Being *towards* this Being" (SZ 12). This relationship is expressed as an understanding. This relationship "in turn means that Dasein understands itself in its Being in some way and with some explicitness" (SZ 12, JS 10). We recognize the characterization of the pre-understanding: that it is vague, but always there. Such a "there" is meant by the term "Dasein", as man is to be called Dasein insofar as he is considered as the location of Being. Later, in a more ambiguous fashion, Heidegger describes this relationship as man being "used", "required", or "called upon" by Being. But such phrases do not refer to anything essentially different from what is denoted by "Dasein". They indicate that we can learn something about Being primarily by making ourselves transparent as "Dasein". Learning something about Being is to move from one kind of "understanding" to another. "Understanding" is Dasein's way of Being, its relation to Being. But, as we shall see, there are modes, or modifications, of that understanding.

The idea of transcendental phenomenology in Being and Time

Fundamental ontology as hermeneutical phenomenology

So far, we have seen that fundamental ontology "grounds" metaphysical ontologies, and that it does so by showing that Dasein always and in every instance "understands" Being. To say that Dasein essentially "understands" Being, that understanding is its basic structure, is to say that it is hermeneutical.

The word "hermeneutical" stems from *hermeneuein,* which simultaneously means to transmit (as Hermes the messenger does), to interpret (as the *hermeneutes,* the priests at the oracle of Delphi do), and to translate (as both Hermes and the *hermeneutes* do). The word has been used in legal and theological traditions to signify theories of textual interpretation (legal or Biblical). Dilthey appropriated this term to encompass the philosophy of the human sciences (*Geisteswissenschaften*) in general. These sciences "interpret" and "understand", whereas the other sciences "know" and "explain".

Heidegger's use of the word "hermeneutics" removes the term from these contexts, from the context of the human sciences. The inquiry into the relationship between Dasein and Being is "hermeneutical" in a new way. "Phenomemology of Dasein is *hermeneutics,*" Heidegger says, "in the original signification of that word, which designates the work of interpretation" (SZ 37, JS 33). Interpretation of what? First of all, the interpretation of Dasein by itself. Dilthey had said "*das Leben legt sich selber aus*" similarly, Heidegger writes, "Self-interpretation belongs to the Being of Dasein" (SZ 312, JS 288). Understanding and interpret-ing are ways of Dasein's Being. Thus: "Every question that is ontologically explicit about Dasein's Being is already prepared by Dasein's own way of Being" (SZ 312).

From this originary hermeneutical layer—Dasein always interprets itself, spells itself out—other modes of hermeneutics are derived. That is, because Dasein always already interprets itself, it can also understand things other than itself, regions of Being other than its own. In this second more derivative sense hermeneutics "works out the conditions of the possibility of every ontological investigation" (SZ 37, JS 33). This simply means, as we have already seen, that the basic structure of Dasein is such as to allow it to distinguish various metaphysical ontologies (such as the "ontology of nature" that is foundational for modern physics in Heidegger's reading of Kant). Finally, hermeneutics designates the description of the structural elements of Dasein, i.e. the existential analytic itself. It is, of course, the first sense that is fundamental. Only because Dasein always already interprets itself can there be a history of such self-spelling-out, as well as a discipline—historiography—of that historical self-interpretation. And only because Dasein interprets itself historically can "hermeneutics" finally designate "the methodology of the historical humanistic disciplines" (SZ 38, JS 33).[34]

These brief remarks are merely intended to emphasize that the word "funda-mental" in the title "fundamental ontology" refers to an event, a self-interpretation or self-articulation of Dasein, which makes clear the difference between "fundamental" and "foundational" ontology. "Foundational" ontologies provide what Heidegger called the "genealogy" of Being's possible ways of con-stituting regions for scientific investigation; but "fundamental" means "interpretive", an act of Dasein, the modalities of its self-understanding.

In this sense then, fundamental ontology is hermeneutics in the originary (*oriri*) way, it is the source for all other modes of understanding and eventually of knowledge. Rational knowledge (and this has often been misunderstood as if Heidegger meant to stand up as a declared enemy of reason) is two steps from that which is "originary". The first step is Dasein's self-interpretation (by which it can be said to "be ontological" [SZ 12, JS 10]). The second is an explicit interpreta-tion or understanding of Dasein, things, or other persons. From these two steps, "theoretical cognition" becomes possible, which Heidegger says "reaches far too short a way" (SZ 134, MR 173) compared with originary understanding.

This hermeneutics, finally, is phenomenological. Fundamental ontology is phenomenology. From what has been said of the retrieval, i.e. the subject matter of *Being and Time*—namely the meaning of Being and of hermeneutics—it should be clear that "phenomenology" here does not aim at the intuition of essences, does not bracket existence, does not describe intentional acts. In other words, it does not explore consciousness. Let me illustrate Heidegger's diver-gence from previous conceptions of phenomenology by examining several texts. Heidegger has already argued that Plato and Aristotle "wrested...[the question of Being] from the phenomena" (SZ 2, MR 21). Later, he will be more explicit:

Husserl watched me in a generous fashion, but at the bottom in dis-
agreement ... I learned one thing ... What occurs for the phenomenology
of the acts of consciousness as the self-manifestation of phenomena has
been thought of more originally by Aristotle and all Greek thinking and
existing as *aletheia.*

(SD 87/79)

In *Being and Time, aletheia* designates precisely the understanding, the interpretation by which Dasein opens up its own realm, a world in which things may show
themselves. Hence the polemical question: "Whence and how is it determined
what must be experienced as 'the things themselves' in accordance with the principle of phenomenology? Is it consciousness and its objectivity or is it the Being
of beings in its unconcealedness and concealment?" (ibid.).

Thus, "the very issue" of phenomenology as practiced by Heidegger since
Being and Time is "Being itself". As the title of Section 7 indicates ("The
Phenomenological Method of the Investigation" [SZ 27, JS 23]), "phenomenology" designates a method. It is that method that lets an investigation be rooted not
in a "dogma", but in the "matters themselves". In this case fundamental ontology
is phenomenological in its method because its inquiry is rooted in "Being itself" as
disclosed by Dasein. Thus, instead of a "pre-given discipline" (SZ 27), "a philosophical discipline may be developed eventually only from the objective necessity
of definite questions and procedures demanded by 'the things themselves'" (ibid.).

In this section, Heidegger simply wants to work out a preliminary conception
of phenomenology (SZ 34). Why "preliminary"? Because phenomenology as a
discipline would eventually result from the research itself. In other words, the traditional relation between the method of an investigation and its subject matter is
reversed. We first look at the subject matter, the thing itself, which is Dasein's
openness to Being; and then the method will progressively take shape.

It is clear that *Being and Time* was possible only on the basis of Husserl's
work.[35] Heidegger says this in a footnote to *Being and Time:* "The following
investigations would not have been possible without the foundation laid by
Edmund Husserl" (SZ 38, JS 34). At the same time, on this page, the opposition is
stated broadly: phenomenology is not to be practiced "in its *actuality* as a philosophical 'movement'" since "[h]igher than actuality stands *possibility*" (ibid.).
And in *On the Way to Language* he says: "I dedicated *Being and Time,* which
appeared in 1927, to Husserl, because phenomenology presented us with possibilities of a way" (US 92/6). This clearly means, on the basis of what has been
said about the philosophy of subjectivity above, that Heidegger rejects the path of
the transcendental ego. Rather than starting from such a point of departure, he
begins with everyday existence, with "factical life".[36] The claim, now, is that this
starting point in actual life is radical phenomenology as opposed to a precon-
ceived discipline, a movement, and finally, yet another dogma.

Stated otherwise, phenomenology is already dogmatic if it claims to set out
with intuition, *Anschauung,* because then phenomenological description can only

be of things that are objects. But if the first "thing" to be described is everyday life, then we place ourselves on a level prior to subject–object dichotomies, which are the most dogmatic in modern philosophy. Transcendental phenomenology recedes, steps back, from things conceived of as objects, to the self conceived of as subject. Such a procedure misses life in its factual determinations—it misses the "origin", *Ursprung,* from which thinking can arise at all. This is what is meant by the substitution of "understanding" for "intuition". And this is what is meant by Heidegger's reversal of phenomenology: it proceeds not by bracketing life, but *as* a "hermeneutics of facticity".

Hermeneutical phenomenology thus means interpreting, translating, and transmitting the message of "Being itself" in such a way that this Being itself becomes apparent in Dasein's everyday life. One way of indicating the opposition to Husserl would be to stress, as Heidegger does time and again, the opposition between seeing and hearing. In hermeneutical phenomenology, it is hearing that has to be learned, whereas in the phenomenology of consciousness, the *Wesensschau* is the goal. This point is illustrated in another passage from "A Dialogue on Language":

> The expression "hermeneutic" derives from the Greek verb *hermeneuein.* That verb is related to the noun *hermeneus,* which is related to the name of the god Hermes. Hermes is the divine messenger. He brings the message of destiny. *Hermeneuein* is that exposition which brings tidings because it can *listen* to a message.
>
> (US 122/29)

"Understanding" designates something that one hears; "intuiting" and, still more, "ideation", designate something that one sees. Thus, phenomenology becomes hermeneutical when its starting point is Dasein insofar as it understands Being— not the transcendental ego, but factual existence. *The ear is attuned to time.* And yet, this hermeneutical phenomenology is transcendental, but in another way. This is what we must now consider.

The transcendentalism of Being and Time

The step backward from foundational ontology to fundamental ontology appeared as a move from one condition of possibility to another. Foundational ontology makes sciences possible. But fundamental ontology makes foundational ontologies possible. We must also remember that this two-step movement brought together two heterogeneous movements. The first step backward is metaphysical, the second is phenomenological. Or, the first step is transcendental in the sense of assigning regions to cognition, while the second step is transcendental in assigning modes to understanding. What Heidegger calls "transcendental knowledge" in *Being and Time* is the question of what Being means for Dasein, and for Dasein in its everyday processes. Thus, the notion of transcendentalism in *Being and Time* is new.[37]

When Heidegger says that Being is *"the transcendens pure and simple [das transcendens schlechthin]"* (SZ 38, JS 33–4), he means that it is the movement by which Dasein steps beyond itself, is in the world, is "ek-static", stands out of itself. It is therefore in relation to everyday existence that we have to understand statements such as: "Every disclosure of Being as the *transcendens* is *transcendental* knowledge. *Phenomenological truth (disclosedness of Being) is veritas transcendentalis*" (SZ 38, JS 34).

In a short essay, "On the Essence of Ground", written in 1928, Heidegger articulates the sense in which fundamental ontology is "transcendental":

> Transcendence means stepping beyond (surpassing) [*Übersteig*]. One has to call "transcendent" or "transcending" that which so "steps beyond". This stepping beyond, understood as a happening, belongs properly to one being. Formally, this stepping-beyond can be construed as a "relationship" that stretches "from" something "to" something. Furthermore, to stepping-beyond belongs that towards which this stepping-beyond leads, and which is usually, but improperly, called "the transcendent". And finally, "something" is always gone beyond (surpassed) in such stepping-beyond.
>
> (WG 33/35/107)

This text indicates the three elements of "transcendence" understood as an event, as a happening. We remember that the event-character of transcendence was present from the very beginning of *Being and Time,* and that "fundamental ontology" is "fundamental" in the sense of the event of founding—e.g. a city, etc. Likewise, Dasein was described as a relationship. Dasein is such that "in its Being it has a relationship" (SZ 12) to Being itself. This relationship and this event-character are now technically called "transcendence". It should be quite clear in what sense Heidegger's transcendentalism is novel: the very Being of Dasein is to step beyond itself, to project itself into what will be called its "world". "Every disclosure of Being as the *transcendens* is *transcendental* knowledge" (SZ 38, JS 34).

"Transcendental", then, still means *stepping back* towards an originary condition of possibility. This condition of possibility, however, is to be understood as "fundamental" in the sense of an event, not as "foundational". "Transcendental" then means *stepping beyond,* which here designates nothing other than the "preliminary understanding" of Being, the vague familiarity, or, what Heidegger also calls the "pre-ontological" understanding of Being (Figure 2.1).[38]

The first of these two senses indicates the method of *Being and Time,* the second indicates the "nature", the "constitution" of Dasein.[39] But the method can only be transcendental because the phenomenologist can trace *back* the movement of springing *forth.*

This transcending, understood as Dasein's stepping beyond itself, is called at times "ontological" and at times "pre-ontological": "Dasein is ontically distinctive in that it *is* ontological" (SZ 12, MR 32); and, "a pre-ontological Being belongs to Dasein as its ontic constitution" (SZ 17, JS 15). The everyday transcendence must

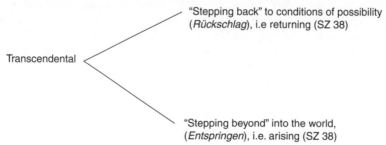

Figure 2.1 The meanings of "transcendental" according to Heidegger

appear as "pre-ontological" when "ontology" or "fundamental ontology" designates the task of making explicit what is always already occurring in Dasein. That
same transcendence must be called "ontological" when "ontology" designates that
hidden structure to be made explicit.

If we keep in mind that "ontic" is the equivalent of "empirical" in Heidegger,
we understand a phrase such as: "Dasein is ontically 'closest' to itself, while
ontologically farthest away; but pre-ontologically it is surely not foreign to
itself" (SZ 16). Stated otherwise: empirically we are always closest to ourselves,
to our own movement of transcendence, since we are that movement. But thematically—ontologically—this is most hidden, precisely because it is so close.
Finally, pre-ontologically, i.e. as inarticulate knowledge of our own transcendence, this movement, this happening, is surely not something foreign. For
Heidegger, *"the ontic depiction of beings within the world"* (SZ 64, MR 92) is
opposed to the *"ontological interpretation of the Being of these beings"* (SZ 64,
JS 60). Thus, there is an opposition between "descriptive" and "transcendental"
phenomenology.

"Transcendental" primarily designates the pre-ontological self-surpassing of
Dasein; only secondarily is it the title for the method of stepping back from the
world, or from everyday life, to the origin in Dasein of that movement of self-
surpassing.

How then, is fundamental ontology "transcendental phenomenology"? It is so
in the sense that the "original" concept of "phenomenon" designates precisely
that "origin", that springing-forth or disclosure, which always occurs in daily life.

In Section 7A, Heidegger identifies three notions of "phenomenon": the formal, the common, and the properly phenomenological. In each of these cases,
"phenomenon" designates "that which shows itself in itself [*das Sich-an-ihm-
selbst-zeigende*]" (SZ 31, MR 54). The "merely formal" notion of phenomenon is
that which is operative in the transcendental tradition prior to its modification by
Heidegger. Thus, the merely formal phenomenon designates that which shows
itself to the transcendental subject, to the worldless ego that knows of no everyday
life. This "formal" notion of phenomenon designates the content of both Kant's
"empirical 'intuition'" (SZ 31, MR 54) and Husserl's correlate of the intentional
subject. The "common" or "vulgar" notion of phenomenon is later equated with

Table 2.3 Transcendental phenomenology as practiced by Heidegger in *Being and Time*

Notion of "phenomenon"	Being as "appearing to"
• Formal	• Transcendental subject
• Common	• Average everydayness
• Phenomenological	• Dasein

"pre-ontological". In the context of one particular phenomenon, namely conscience, we read: "Even the common [vulgar] experience of conscience must somehow—pre-ontologically—reach this phenomenon" (SZ 289, MR 336). Thus, the second notion of "phenomenon" must be equated with a non-thematic, "vague" encounter, i.e. these are the phenomena of what Heidegger will call *das Man,* average everydayness. It is that which "already shows itself in appearances prior to and always accompanying what we commonly understand as phenomena, though unthematically" (SZ 31, JS 27–8.). Finally, the "phenomenological" concept of phenomenon—which in the *Basic Problems of Phenomenology* Heidegger calls "scientific"—is that which was referred to in the very first section of the book. Plato and Aristotle "wrested from phenomena, in the highest exertion of thinking" (SZ 2, JS 1), the meaning of Being. Transcendental phenomenology as practiced by Heidegger in *Being and Time* is the effort of wresting the phenomenon of Being from Dasein, thus Table 2.3.

If it is possible explicitly to retrieve "what already shows itself in appearances prior to and always accompanying what we know"—and do—then the phenomenon so retrieved is "to be called 'phenomenon' in a distinctive sense [*in einem ausgezeichneten Sinn*]" (SZ 35, JS 31). The subject matter of transcendental phenomenology is then:

> … something that does *not* show itself at first and for the most part, something that is *concealed,* in contrast to what at first and for the most part does show itself. But at the same time it is something that essentially belongs to what at first and for the most part shows itself, indeed in such a way that it constitutes its meaning and ground.
>
> (SZ 35, JS 31)

It has thus been shown how fundamental ontology is hermeneutical phenomenology: *"Ontology is possible only as phenomenology"* (SZ 35, JS 31). It has also been shown how this phenomenological ontology is transcendental: it is entirely a phenomenology of Dasein as stepping beyond itself, as transcending itself (*transcendere*) and moving towards the world. As such a phenomenology, it steps back to the "relationship" that is Dasein's Being or the condition of possibility of all modes of Being-in-the-world.

The working-out of the transcendental-phenomenological method

It should now be clear that to adopt phenomenology as a discipline, a movement, or an established method, would be dogmatic. Methodologically, that leaves *Being and Time* in a precarious position since a central object of inquiry for the book is the method of inquiry itself.[40]

The "question of Being" is retrieved in *Being and Time* via "an analytic of Dasein" (of which nothing has yet been said). But the analytic of Dasein can only function as revelatory of the phenomenon, as it has just been worked out, if that analytic is guided by some "idea", some thread by which we hope eventually to be led to the meaning of Being. This guiding idea, Heidegger says, is the "idea of existence", what he calls "the Being of Dasein", i.e. *"the 'essence' of Dasein lies in its existence [Das Wesen des Daseins liegt in seiner Existenz]"* (SZ 42). Heidegger is quick to add that this has nothing to do with the metaphysical distinction between essence and existence ("ek-sistere", ecstasis).

What is important here is that "the idea of existence", i.e. of Dasein's Being as ek-static, guides transcendental phenomenology, insofar as it takes its point of departure in the analytic. The "idea of existence" is the anticipated guiding light. Since "ek-sistence" means nothing other than the very movement of transcendence, this guiding idea speaks of what is indeed most basic in Dasein.

This guiding idea is one that determines Dasein as a whole. It designates Dasein's whole constitution, its structured totality. "Dasein's whole constitution itself is not simple in its unity, but shows a structural articulation" (SZ 200, MR 244). This structural articulation is at times called "Being-in-the-world", at other times "the idea of existence", and again "care".

Transcendental phenomenology is thus a transition from the singular to the general to a whole. In this sense, the existential analytic does not deal with ontic experiences—it is neutral. The idea of existence is the idea of a structural, "neutral" totality. Concerning this, Heidegger says:

> We come to terms with the question of existence always only through existence itself. We shall call *this* kind of understanding of itself *existentiell* understanding. The question of existence is an ontic "affair" of Dasein. For this the theoretical transparency [perspicuity] of the ontological structure of existence is not necessary. The question of structure aims at the analysis of what constitutes existence. We shall call the coherence of these structures *existentiality.* Its analysis does not have the character of an *existentiell* understanding but rather an *existential* one.
>
> (SZ 12, JS 10f)

It is clear that this distinction between "existentiell" and "existential" is one between singular and universal, particular and general. The structural analysis will eventually indicate some ontic, concrete, possibilities, but this is not its goal. Structural analysis is meant to remain "neutral" with regard to these possibilities.

Thus, when showing one such structure, that of "death", Heidegger will say:

> The fact that in an existential analysis of death, existentiell possibili-
> ties...are consonant with it [*anklingen*], is implied essentially by all
> ontological investigation. All the more explicitly must the existential def-
> inition of concepts be unaccompanied by any existentiell commitments.
>
> (SZ 248, MR 293)

Thus, "existence" as the guiding idea for the analysis indicates the movement from the particular (experience) to the whole of Dasein's structures. The methodological expression of this direction given to the analysis is the thesis of neutrality, of non-commitment. It is the "direction" earlier called the "return" (*Zurückgang*). It is balanced, or complemented, by another direction taken later in the book, which follows the "arising" (*Entspringen*).

The first direction, that of the "return", is from phenomena to their *Ursprung* (origin), springing-forth into the whole of Dasein. Here the phrase "the whole of Dasein" designates a structural, formal, neutral, and non-empirical whole,[41] and this is the reason for Heidegger's repeated insistence that Dasein is not something "objectively present".[42] He says: "[t]his being does not have, and never has, the kind of being of what is merely objectively present in the world. Thus, it is also not to be thematically found in the manner of coming across something objectively present" (SZ 43, JS 40). To interpret Dasein as something objectively present—as both the Greeks and the moderns do—amounts to a "reification" (SZ 437, JS 397) of Dasein. The *fact* that Dasein has historically been so reified, must give rise to a questioning of its origin: "What does this 'reifying' signify? Where does it arise?" (SZ 437, MR 487). The direction of the "return" follows this backward arising to the *Ursprung,* which, at least in *Being and Time,* is Dasein as a structural whole.

The second direction, that of the "arising", is an unfolding of the so-called "existential determinations", such as care, dread, being-towards-death, etc., out of their originative source. To constantly tie these determinations back to their source, to the "formal idea of existence", is to understand them in their unity: "The existential characteristics are not pieces belonging to something composite, one of which might sometimes be missing; but there is woven together in them an originative context" (SZ 191, MR 235).[43] Thus, when Heidegger begins to unfold this complex fabric of existential determinations, he can say: "The preliminary glance which we gave to the whole of this phenomenon (of Being-in-the-world) in the beginning has now lost the emptiness of our first general sketch of it" (SZ 180, MR 225). This first general sketch is summarized with the phrase "Dasein exists" (SZ 53, JS 49). In the unfolding of the existential determinations, the formal idea of existence is made progressively more concrete.

The reason this is addressed within the context of methodological considerations is that the understanding of totality undergoes a change. In what we have called the "return", the structural unity of all *existentialia* is called the whole, the

81

totality. But with the unfolding of these *existentialia,* the wholeness comes to be located in what Heidegger calls a "concrete phenomenon". The entire second division of *Being and Time* constitutes a quest for some phenomenal testimony that "Dasein can be whole". Now it is authentic existence that is called "whole". The wholeness of Dasein no longer lies in its fundamental structure, but in one possibility, one modification of it. Authenticity and inauthenticity are, to be sure, lasting possibilities—they are structural elements. But among the many structural elements, one now bears witness to wholeness, i.e. to the "authentic Being of Dasein" (SZ 311, JS 287).

This change in the direction indicates the main reason the method of *Being and Time* is so difficult to lay bare. Heidegger goes back and forth between two understandings of wholeness. At one point, wholeness is the formal-transcendental, fundamental structure. At another point, it is "a possibility", which he will later call "*Ganz-sein-können* [potentiality-for-Being-a-whole]" (SZ 305, MR 353).

This change also explains why, as the book progresses, we are told that we have to move further and further back to seize what is *ursprünglich* (originative) in Dasein. First it is Being-in-the-world in general, and with respect to this, Heidegger states that the initial abstractness has now been overcome. But then he asks, "How is the totality of that structural whole...to be defined in an existential-ontological manner?" (SZ 181, MR 225). Heidegger answers:

> To put it negatively, it is beyond question that the totality of the structural whole is not to be reached by building it up out of elements. For this we would need an architect's plan. The Being of Dasein, upon which the structural whole as such is ontologically supported, becomes accessible to us when we look all the way *through* this whole *to a single* originative and unitary phenomenon [*ein ursprünglich einheitliches Phänomen*], which is already in this whole in such a way that it provides the ontological foundation for each structural element in its structural possibility.
>
> (SZ 181, MR 226)

This one originative phenomenon will be called dread (*Angst*). But is it not strange that the wholeness should be ascribed to one phenomenon, and that this should even be said to provide the ontological foundation for all other structural elements or *existentialia?*

The question of method rebounds: dread is still not really originative. A yet more originative whole is care. At the beginning of the second division, doubts arise as to the originary character of the first division altogether: "Are we entitled to the claim that in characterizing Dasein ontologically as 'care' we have reached the *originary* interpretation of this Being?" (SZ 231, JS 214). No, we are told, because the phenomenon of care has not yet appeared in its ontological meaning (SZ 303). A new level of origination is gained with the discovery that "time is the ontological meaning of care" (SZ 323).[44] As we know, authenticity becomes thinkable only in the context of time—as authentic temporality.

Table 2.4 Heidegger's development of the transcendental phenomenological method

Formal idea of existence	Inauthenticity	Authenticity
• Being-in-the-world	• The "they"	• Anticipation of death
• Care	• Falling	• Resolution
• Being-towards-death	• Everydayness	• Anticipatory resolution
• Originary temporality	• Objective presence	• Authentic retrieve

Heidegger's analysis thus constantly follows the same course: the course of the *"Entspringen"* or *"Ursprung"* of phenomena *out of* their transcendental structural ground.[45] In one direction he traces back the *Selbstentfaltung,* the process of unfolding itself, which belongs to Dasein—this is why its Being is called ek-sistence. In this direction, he discovers the *formal whole.* In the other direction, following the self-unfolding, he discovers *phenomenal wholes.* Only in this back-and-forth is the analysis of Dasein progressively deepened.

In this back-and-forth, we observe the working-out of the method. It is somewhat experimental, since Heidegger would consider it dogmatic to set out with a method completely worked out in advance. Again and again this deepening thematizes the same elements, yet under different names. We can now lay these elements out in three series (Table 2.4).

In the first column, Heidegger positions himself within the transcendental tradition. In the second column, he renovates that tradition starting from everyday life. In the third, as we shall see, he opens new possibilities for practical philosophy.[46]

The general structure of the understanding of Being

We now know why Da-sein is the exemplary being for the retrieval of the "question of Being". The methodological considerations revealed a series of terms that all refer to the connectedness between Dasein and its world. These terms all point to a "stepping-beyond", literally to Dasein as "out-standing"—they are ek-sistence (Dasein's Being is existence), ek-stasis (Dasein's Being is to stand out of itself in past, present, and future), trans-cendence (Dasein's Being is to surpass itself), and finally "Being-in-the-world". This latter term is the guiding concept by which Heidegger battles the solipsism of the Kantian "subject". It is the concept that most clearly and explicitly indicates the way in which Dasein is a formal, structural whole. It renders more concrete the vague familiarity that we have with Being, i.e. "Being-in-the-world" reveals the first layer of the structure of *Seinsverständnis* (understanding of Being).

In this effort to grasp the general structure of *Seinsverständnis,* we will have to proceed in a threefold manner: existence, ek-stasis, and transcendence are all titles for the "understanding of Being". It is because we "exist ecstatically", i.e. because we always "transcend" ourselves, that we always already understand Being. To make this understanding explicit, it is first necessary to show how we misunderstand our own general structure as long as we do not speak of it in terms of the "world". In the second step, the totality of our constitution is questioned in two more "rounds" of deepening investigation. The "ground" of Being-in-the-world is care, Heidegger will say, and the "meaning" of care is temporality. Hence, this constitutes the threefold approach to the general structure of the understanding of Being: Being-in-the-world, care, and temporality. All three titles speak of the whole of Dasein, but not in the same way. And all three speak of Dasein as the locus where the question of Being can be retrieved, but again, not in the same way. Due to the cyclical procedure of *Being and Time,* the same themes, e.g. "project" and "thrownness", will reappear at each of the different layers of "originariness".

Being-in-the-world

In order to clarify these three approaches, the first of which is the explication of Being-in-the-world, I will quote the relevant texts from the second and third approach. The "Being of Dasein" is called existence in "formal" terms, but care (*Sorge*) in "concrete" terms. When Heidegger introduces this chief *existentiale,* care, he says that it unfolds the constitutive moments of Being-in-the-world, namely "world, Being-in, and self" (SZ 190, JS 178). Thus, Being-in-the-world is concretized for the first time with the title "care". It is concretized the second time with the title "temporality": "Only through the rootedness of Dasein in temporality can we get an insight into the existential *possibility* of that phenomena which, at the beginning of our analytic of Dasein, we have designated as its basic constitution, namely *Being-in-the-world*" (SZ 351). This quote indicates that the inquiry will lead toward an increasingly originary grasp of the wholeness of Dasein, that is, its formal-transcendental wholeness. The existentiell wholeness, authenticity, will have to be considered thereafter.

The task that Heidegger undertakes in the Section "Being-in-the-world" is "to bring into relief phenomenally the unitary originary structure of Dasein's Being, in terms of which its possibilities and its ways of 'Being' may be ontologically determined" (SZ 130). All the concepts mentioned earlier, existence, ek-stasis, and transcendence, point to what Heidegger now calls the structure of "Being-in". Indeed, in everyday life, we appear as "disclosed" (*erschlossen*). "In the expression 'there' [Da-] we have in view this essential disclosedness" (SZ 132, MR 171). This does not mean that we constantly experience ourselves explicitly as so "disclosed" towards a world; on the contrary, "proximally and for the most part" this disclosedness remains unthematized.

"Being-in" or "Being disclosed" designate, at this level, Dasein's Being: "Dasein is its disclosedness" (SZ 133, JS 171). Thus Being-disclosed in a world is assuredly a total structure, a name for the fabric of *existentialia*. At the same time, it is a complex and structured whole. As Heidegger writes, "In *attunement* and *understanding,* we see the two constitutive ways of Being the 'there'... Attunement and understanding are characterized equioriginarily by *speech*" (SZ 133).[47] Let us first look at these three structural components of Being-in-the-world.

Attunement, understanding, speech[48]

Attunement is one aspect of Being-in-the-world. We are always attuned to what surrounds us. To speak of attunement (*Befindlichkeit, Stimmung*) is, of course, a clear way to discard rationalist approaches to Being-in-the-world. In everyday life, what we experience is that we are "in" the world according to moods, states of mind, which disclose and foreclose certain domains of things. Attunement is a mode of disclosedness, the most current one with which everyone is familiar, and one that can never be—despite most philosophers' exhortations—overcome. In "On the Essence of Ground" Heidegger speaks on this matter:

> If the world did not appear, or at least dawn, in Dasein's preoccupation with being, it could not be attuned to beings; nor, then, could Dasein be, for example, embedded or stifled or permeated by beings. For it would lack the necessary "leeway"... it is only as being-in-the-world that Dasein is preoccupied with beings. Dasein grounds, or establishes, world insofar as it grounds itself in the midst of beings.
>
> (WG 62/109–11/128)

The second element, understanding, is a more active determination: it is a capacity (*können*). To understand one's world is to find one's ways about it. This finding-one's-ways-around is always attuned in such a way that understanding is but another facet of Dasein's grounding in the midst of beings. The difference lies mainly in this capacity, as will appear later, since understanding is, in the end, the capacity for projection. Attunement and understanding can be grasped in their particularity only in contrast to one another. We have no real hold on attunement. But understanding is the ground for having a hold on the world, a ground for projects. Furthermore, understanding discloses the world, "our" world, as a whole within which we know that we can undertake things. Attunement discloses Dasein primarily as "brought before itself"—but in such a way that it finds itself exposed [to itself].[49] Attunement "comes upon" one.[50]

Attunement and understanding must be grasped together in order to see the double strategy that we have already traced through the exemplary character of Dasein. First, there is a strategy of wholeness, i.e. understanding, and second, a strategy of differentiation, i.e. attunement. When Heidegger begins his analysis of world as context of equipment (*Zeugzusammenhang*), the world appears immediately as differentiated. Within everyday life, the instruments that we use refer to one another, to gestures and use. Thus, proximally and for the most part, the world appears as a *"referential totality [Verweisungszusammenhang]"* (SZ 82, JS 77). It is understanding that seizes this referential totality; it seizes the world as "my" world and as a whole. "The context of equipment is lit up, not as something never seen before, but as totality sighted beforehand in circumspection (*Umsicht*). With this totality, the world announces itself" (SZ 75, MR 105). I mention this referential totality that is our daily world here only to indicate the structural whole that

Heidegger tries to grasp through the notions of attunement and understanding. But, of course, the context of equipment is only one—the most ordinary—instance of such wholeness. What is important to see, in this thematic of the world, is that the world is not merely a horizon, but "the...whole of *possible* interconnection" of beings (SZ 144, MR 184). One such possible interconnection is the context of equipment.

More explicit than attunement is understanding, and more explicit than understanding is the third structure of Being-in-the-world, speech. What this "*existentiale*" does, specifically, is to render explicit the referential connectedness of what we deal with in everyday life; it "articulates" understanding. "The fundamental '*existentialia*' which constitute the disclosedness of Being-in-the-world, are attunement and understanding... *Existentially equi-originary with attunement and understanding is speech*" (SZ 160–1). Speech (*Rede*) is called the "existential-ontological foundation of language" (SZ 161, JS 150), that is, speech is not primarily the sequence of statements that we utter; it is the condition that makes such utterances possible.[51] Even one who does not speak is determined in his Being-in-the-world by this condition. Heidegger puts the three "originary" *existentialia* together in one brief phrase: "the understandability, which is attuned, of Being-in-the-world *expresses itself as speech*" (SZ 161). But there is not a sequence, as in the Aristotelian theory of abstraction, from something sensible to something intelligible and then to something uttered; rather "speech is existentially equioriginary with attunement and understanding" (SZ 161, JS 150). These three are called "originary" or "fundamental" *existentialia* since they structure our Being-in-the-world.[52] One can also say that they are the three basic modes—always given simultaneously—of being "there", Da-sein.

Thrownness and projection

The three *existentialia* just mentioned determine the whole of Dasein and show its inner structuring, its inner articulateness. But the question—since the entire analysis of Dasein only serves to retrieve the question of Being—is: how is the understanding of Being "always already" operative in daily life? How is it so that the total fabric of existential determinations *are* our "understanding of Being"? Attunement, understanding, and speech are one "cut", so to speak, of Being-in-the-world. This cut intersects with another cut, according to which Dasein finds itself both "thrown into" the world and "throwing itself forward", pro-jecting, towards the world. Thrownness and projection (*Geworfenheit* and *Entwurf*) further articulate the entirety of Dasein, but from another angle.

Attunement, understanding, and speech do not imply that Dasein can be affected by things or even threatened. Thus, this new approach starts with "*Betroffenwerden*", "*Bedrohbarkeit*", and, in more general, neutral terms, with "*Angänglichkeit*", with the fact that something can matter to us. If things can "matter" at all for Dasein, this capacity is rooted in attunement, which presupposes that the world has already been disclosed. Only something that is in the attunement of fearing (for instance) can discover that things in the world are threatening. And then, "Dasein's openness to the world is existentially constituted by the mood of an attunement" (SZ 137).[53]

Moods, then, indicate that we are "thrown into the world", at the disposal of events. "[F]rom the ontological point of view we must as a general principle leave the primary discovery of the world to 'bare mood'" (SZ 138, MR 177). Mood indicates a way we are "angegangen" by things, the way they "matter" to us. "We are never free of moods. Ontologically, we thus obtain as the first essential characteristic of attunement that attunement discloses Dasein's throwness" (SZ 136, MR 175). Moods only indicate that we are put at the disposal of things around us. Heidegger adds that in everyday life, the first reaction to such confrontation is evasion: "[attunement] disclose[s] Dasein in [its] throwness, and—proximally and for the most part—in the manner of an evasive turning-away" (SZ 136, MR 175). Moods "come over us", and thus they indicate that, first of all, the world comes over us.

Moods indicate not only that we are thrown into a world, but that we are and have to be: "Being has become manifest as a burden...In having a mood, Dasein is always disclosed moodwise as that one being to which Dasein has been delivered over in its Being—Dasein as that one Being which, in existing, it has to be" (SZ 134, MR 173). This task, to be, which Dasein understands in throwness, indicates the second facet, project.

Moods and attunement reveal possibilities for Dasein.

> Dasein, as essentially attuned, has already got itself into definite possibilities as the potential-for-Being which it is, it has let such possibilities pass by; it is constantly waiving the possibilities of its Being, or else it seizes upon them and makes mistakes. But this means: Dasein is a potential Being that has been delivered over to itself; it is a thrown potentiality, through and through.
>
> (SZ 144)[54]

This throwness sheds a new light on the three previous existentialia. We have seen that throwness is discovered only through moods and attunement. We have also seen that in throwness we understand that Being is a task. With regard to speech, its very possibility, we are told, resides in throwness. "Speech is existentially language [Sprache] because [Dasein] has as its kind of Being, Being-in-the-world—a Being which has been thrown and submitted to the 'world'" (SZ 161).

It is the potentiality-for, the possibility-to, that reveals the correlative existentiale: throwness indicates project. Together they show, once again, that being-in-the-world is a whole, but a differentiated whole. Throwness and project articulate Dasein again in its entirety. In fact, all three previous existentialia— attunement, understanding, and speech—have revealed Dasein as potentiality. Thus, they help in opposing an understanding of Dasein as an objectively present being. "Dasein is not objectively present, something which possesses its potentiality for something by way of an extra; it is primarily Being-possible" (SZ 143). *Möglichkeit,* which stems from *mögen, vermögen, machen* (being able to undertake something), is Dasein's own Being: "Dasein is in every case what it can be" (SZ 143, MR 183).

Needless to say, this concept of potentiality must be distinguished from the logical use of "possibility"—for instance, in transcendental logic, as one of the categories of modality.[55] Only for things objectively present (*Vorhanden*) can possibility and actuality be opposed—not so for Dasein. Therefore, the connotation of "potential" is always present in the talk of *Möglichkeit* in *Being and Time.* Now the transition to "project" becomes clear: "Projection, in throwing, throws before itself the possibility as possibility" (SZ 145, MR 185).[56]

"Projection" designates the way Dasein relates to its Being. Dasein is always more than itself at any given moment, since it holds its own Being before itself as a possibility; it is the potentiality-for-Being. In attunement, understanding, and speech, it "lets be" (ibid.) such possibilities, that is, it makes them its own. With regard to understanding, Heidegger states this in the following way: "Understanding has in itself the existential structure which we call '*projection*'... The character of understanding as projection is constitutive of Being-in-the-world with regard to the disclosedness of the 'there'" (SZ 145, MR 185), i.e. the "there" of Da-sein. Thus, the "there" comes to mean the givenness of a "can-be [*Da eines Seinkönnens*]" (SZ 145).

Now, such projecting has nothing to do with drafting a plan and acting according to it. Rather, it reveals what will later in the book be called the pre-eminence of the future. Projection shows the "fore-having, fore-sight, and fore-conception [Vorhaben, Vorsicht, Vorgriff]" (SZ 150, MR 191) of Dasein's totality as potential. The projected possibility always has the character of wholeness. But this only indicates that the "guiding idea" of the existential analytic now appears in a new light. It is "possibility" that is the impetus for projection; Dasein projects itself towards its possibilities as towards its own possible totality.

Here is one of the *existentialia,* then, that comes to stand for the "guiding idea" of wholeness, i.e. *Möglichkeit.* Besides the progressive deepening whereby this guiding idea receives ever new, more "originary" names, there is something else to keep in mind. In the context of thrownness and projection, "possibility" is guiding. It guides the projection within thrownness, while at the same time guiding the analytic as a whole. In other words, the analytic of Dasein is nothing other than explicit existing. We have seen that this reverberation upon existentiell, concrete, modes of existing becomes thematic only in the analysis of temporality. In the context of authentic temporality, the existential interpretation "must be guided, in every step, by the idea of *existence* ... this means ... that it projects existenti*al* phenomena upon existenti*ell* possibilities, and 'thinks these possibilities through to the end' in an existential manner". If we do this, the working-out of a wholeness that is possible in an existenti*ell* way "will lose the character of an arbitrary construction. It will have become a way of interpreting whereby Dasein is set free *for* its uttermost possibility of existence" (SZ 302–3, MR 350; my emphasis). In other words, existential interpretation yields a liberating power for existentiell possibilities. To philosophize about existence is itself determinate existence.

The hermeneutic of Dasein likewise has the character of projection. One can say that it sets before itself the "idea of wholeness" only because Dasein always sets such an idea before itself. Thus, the existential hermeneutic does what Dasein does: project a form of wholeness. The analytic *develops* the significance of "projecting" that is always operative in Dasein. *Seinsverständnis,* whether implicit in everyday Dasein or explicit in the existential analytic and fundamental ontology, always has the character of projection.

Structurally, projection indicates a whole that Dasein can, of course, never *concretely* achieve. The projecting towards possibilities is always richer than what can in fact be achieved. Since Dasein is "thrown" into a factual world, not everything is possible. By virtue of its own "facticity" many possibilities are taken away from Dasein.[57] Hence the finitude of Dasein's world, i.e. of those possibilities that it can, in fact, seize upon. Thus, thrownness is, so to speak, the principle of limitation upon projection.

"Being-in-the-world" is an altogether preliminary characterization of the constitution of Dasein. It serves as the guiding thread of the analytic. It operates negatively, "prohibitively", since the concept of "Being-in-the-world", as developed in Section 13, is meant to reject from the outset:

- any concept of subject that would be worldless[58]
- any concept of world that would be "theoretical" at its source, as if the world were an object for knowledge.

It also operates positively, since "Being-in" reveals an identity of disclosure that is "always already" (*immer schon*) involved. As shown in Section 41, this preliminary characterization of Dasein's constitution leads to an entire fabric of determinations that are synthesized under the label "care" (*Sorge*). All these analyses that "care" brings together render yet more explicit both the wholeness of this structure and its inner articulation. Therefore, we must now turn to a consideration of "care".

Care

The problematic of care is similar to that of Being-in-the-world. The question is that of the understanding of Being, and why an analytic of Dasein is necessary to "retrieve" the question of Being. The way that this question is answered at the three levels of the general organization of *Being and Time*—Being-in-the-world, care, and temporality—is by showing:

- the wholeness of the structural-functional fabric
- its inner articulation according to concrete possibilities.

The heuristic function of dread (Angst)

"Care" meets these two criteria, but in a more specified manner than "Being-in-the-world". Indeed, "care" is another term for the general, formal, whole of Dasein. But at the same time, there is one particular possibility that directs the inquiry here, i.e. one extreme modification of Being-in-the-world: the world "collapses into itself; the world completely lacks significance" (SZ 186, MR 231). This experience of collapse is one mode of attunement, i.e. being attuned to the world in such a way that it appears threatening—no particular threat is implied here, but the world "as a whole" is a menace. We are attuned by fear to things that threaten us; but the particular threat whereby the world itself, as a whole, loses its significance, is something different—it is dread (*Angst*).

"*That in face of which one has dread is Being-in-the-world as such*" (SZ 186, MR 230). *Angst* "*vor*" is dread in the face of the world and *Angst* "*um*" is dread about Being-in-the-world itself. "Dread individualizes Dasein for its ownmost Being-in-the-world" (SZ 187, MR 232). "Dread takes away from Dasein the possibility of understanding itself, as it falls, in terms of the 'world'" (ibid.). But this extreme possibility is highly revelatory: "*Dread*, as one mode of attunement, discloses the *world as world*" (SZ 186, JS 175).[59] Thus, one particular *existentiale* again reveals the totality of the formal constitution of Dasein. Only because of this revelatory character is dread significant here—not because it fits best into Heidegger's idea of what life should be like.

The way in which dread is related to projection is clear: the world can only "leave me alone" if I always already go out into it. When the world's interconnectedness collapses, this interconnectedness appears as a fabric, which makes clear that "*that which discloses and that which is disclosed are existentially selfsame*" (SZ 188).[60] Dread, then, reveals that Dasein is always ahead of itself in a world. It reveals this Being-ahead, this projectional character, by way of impossibility. To say that Dasein is always "beyond itself" is *not* to say, if we have understood the phenomenon of dread, that Dasein is concerned with one thing or another.

> Dasein is always already *ahead* of itself in its Being. It is always "beyond itself" not as a way of behaving towards other beings which it is *not*, but as Being towards the potentiality-for-Being which it is itself. This structure of Being by which Being is for Dasein itself "an issue", we shall denote as Dasein's "*Being-ahead-of-itself*".
>
> (SZ 191–2, MR 236)

Wholeness and differentiation of "care"

What was first called "Being-in" is now described more strictly as "Being-ahead-of-itself in the world". In other words, this way of speaking includes the wholeness as well as the inner articulateness of Being-in-the-world. Wholeness here means Being-ahead-of-itself in a world, and inner articulateness means

Being-ahead-of-itself,
i.e. "existentiality"

Care

Being-already-in,
i.e. "facticity"

Being-alongside,
i.e. "fallenness"

Figure 2.2 The definition of care

Being-alongside-with-things that are encountered in that world. Later, in the third approach (according to temporality), the Being-ahead will appear as Being-ahead in time. But now, with regard to care, Heidegger says: "The Being of Dasein means 'ahead-of-itself-Being-already-in (the world) as Being-alongside (beings encountered within-the-world)'. This Being fulfills the signification of the term 'care'" (SZ 192, MR 237).

Structurally, care is again differentiated from within, but in a new way. It does not appear according to attunement, understanding, and speech, but as encompassed by what is here called "Being-already-in". Indeed, these three *existentialia*—attunement, understanding, and speech—arose out of an examination of the structure of "Being-in". Care is thus more concrete and wider in formal scope than Being-in-the-world. Besides "Being-in", it stresses two additional features: Being-ahead-of-itself, and Being-alongside. The definition of care can thus be schematized as shown in Figure 2.2.

We have here the three most "encompassing" elements of the formalism in *Being and Time*. I say "formalism" because these three elements designate the formal, not the ontic/existentiell mode of wholeness of Dasein.

Facticity, as being thrown-into-the-world, and fallenness, as being dispersed-among-beings, are familiar titles by now. But what is meant here by "existentiality"? "Care does not characterize just existentiality, as if detached from facticity and fallenness; on the contrary, it embraces the unity of these determinations" (SZ 193). Among these three determinations, existentiality seems to have some kind of pre-eminence. We are told that "care for oneself" (*Selbstsorge*) would be a tautology. "'Care' cannot stand for some special attitude towards the self; for the self has already been characterized ontologically by 'Being-ahead-of-itself'" (SZ 193, MR 237).[61] This is revelatory. We now have a new equation:

existentiality = Being-ahead-of-itself = the self

The text continues: "in this characterization the two other structural moments of care—namely Being-already-in...and Being-alongside...have been *posited simultaneously*" (SZ 193). We thus have a new triangle (Figure 2.3).

91

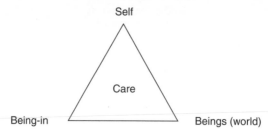

Figure 2.3 Simultaneous positing of the two other structural moments of care, Being-already-in and Being-alongside

Care is now, once again, the most appropriate title—before being replaced by "temporality"—for the structural whole of Dasein. Thus, "[f]rom this signification every tendency of Being which one might have in mind ontically, such as worry or carefreeness, is ruled out" (SZ 192, MR 237). The concept of care, we are told in a footnote "has grown upon the author in connection with his attempts to interpret the Augustinian...anthropology with regard to the fundamental principles reached in the ontology of Aristotle" (SZ 199, MR 492n. vii). The reciprocity of thrownness-projection can be traced back to Augustine. With the Latin word *cura* "what we have in view is one *single* basic state in its essentially twofold structure of thrown projection" (SZ 199, MR 243). In Augustine, being "thrown" is structurally similar to "creation", while "projection" is structurally similar to the restless heart. By calling care an "ontological-a priori generalization", Heidegger also reveals an Aristotelian heritage. This leads Heidegger toward an understanding of "[t]he transcendental 'generality' of the phenomenon of care as well as of all fundamental *existentialia*" (SZ 200, MR 244). Does this amount to Aristotle's very quest for categories?

Since care is merely the formal title of Dasein's totality, it clearly operates on a level prior to the two possible modifications of Dasein, authenticity and inauthenticity. The more detailed implications of the care-structure will appear when we oppose these two modifications concretely.

Methodologically, the care structure has been arrived at through one revelatory *existentiale,* dread. In dread, the totality of the care-structure is given phenomenally.[62] Dread completely discloses what is then diversified through the elements of care. Dread is thus divested of any psychological connotation, and also of any ethical connotation (which it still has in Kierkegaard, from whom Heidegger borrows much of his analysis). Dread is that mood by which man is determined not by particular things, but by a way in which the world as a whole appears—or disappears.

To speak in terms of "care" is Heidegger's way of rephrasing the observation that Dasein is that being in whose being Being itself is always an issue, or always at stake.[63] "Care" means "Being-at-stake", being an issue. Thus the care-structure deals, although rarely explicitly, with the "relatedness" by which the question of Being arose at the beginning of the book. Thus, care concerns more than the Being of man himself, the "self". Care is also care for beings in their totality—it is the unity of this dual "concern" for both other beings and for Dasein itself. Care is that non-cognitive openness to beings by virtue of which they can appear to us at all. This transcendental ground for appearance has to be called "care" in a phenomenology that starts, not with acts of consciousness, but with the phenomenality of everyday life. This phenomenality is itself constituted in such *existentialia* as dread.

"Care", then, is one name for the disclosedness of things. This disclosedness (*Offenbarkeit*) is the happening (*Vollzug*) of Being as it can be described according to the analytic of Dasein. Care is a process, and not one that is accomplished through reason; as an everyday-process it discloses Being itself, it is the name for that disclosure. One could also say that care is the event of transcending. In transcending, self, beings (world), and Being-in are brought together in such a way that Being appears as their unique unfolding, the *transcendens* pure and simple.

This triangle Self/Being-in/beings (world) shows, once again, the tendency, in *Being and Time,* against "decontextualization" of the self (*Entweltlichung*). The self is always already "in"—that is, involved—in its world. This clearly means that we cannot continue speaking of the self as designating a person's inner states or moral-subjective apparatus, and even less a spiritual substance. The self will have to be understood as "Dasein", that is, already engaged in a world. But the pre-eminence of the self—of existentiality—consists in the revelatory function that it has with regard to others. Thus we have to raise in *one* question the problems of the self and of the others.

The self and being-in

The concept of "self" in *Being and Time* answers the question: "Who is this being, Dasein?" (SZ 114). "Dasein is a being which is in each case *I myself*" (SZ 114, MR 150; my emphasis). The concept of self thus points to me, in each case, as the subject matter of the analytic. In that sense, its significance is both ontological and ontic: "This determination [i.e. "self"] *indicates* an *ontologically* constitutive state, but it does no more than indicate it. At the same time this tells us *ontically* (though in a rough and ready fashion) that in each case an 'I'—not Others—is this being" (SZ 114, MR 150). Ontically, the self points to what Heidegger elsewhere calls *Jemeinigkeit,* "mineness". But what is this ontological state that the concept of self merely indicates? It probably results from "Being-in" and "beings" (world).

Heidegger writes: "The 'who' is what maintains itself as something identical throughout changes in its experiences and ways of behavior, and which thereby relates itself to this changing multiplicity" (ibid.). The reference is to Kant's notion of *Beharrlichkeit* (permanence) in the first analogy of experience. The determination "self" only *indicates* an ontological structure because the question and answer "Who?"—"the self" does not yet discard the substantialist misunderstanding. In Heidegger's terms, it is according to this misunderstanding that Dasein becomes at all comparable to something objectively present (*vorhanden*). This misunderstanding is what he discards by opposing the self to the "subject".

In a later development he briefly summarizes the way "self" is operative here, precisely by referring it back to "existing". "The question of the 'who' of Dasein has been answered with the expression 'self'. Dasein's selfhood has been defined formally as a *way of existing,* and therefore not as a being that would be objectively present (present-at-hand)" (SZ 267, MR 312).[64]

Thus, we gather that the self, as an ontological co-constituent, is as much of a process as ek-sisting, trans-cending, etc.—which has a certain number of consequences that bring us to the core of the radical transformation of the understanding of man in *Being and Time.* The transformation is radical because:

- the self is not given
- it is encountered in every-day life (and not in reflection)
- it is modifiable.

These three senses of transformation are elucidated as follows.

THE SELF IS NOT GIVEN

"[W]hat is more indubitable than the givenness of the 'I'?...Perhaps [this givenness] is indeed evident. This insight even affords access to a phenomenological problematic in its own right, which has its significance as 'formal phenomenology of consciousness'" (SZ 115, MR 151). But against the guiding idea of substantiality, Heidegger—as we have seen from the beginning—has worked out a guiding idea of existentiality. This guiding idea is here expressed in the vocabulary later appropriated (and transformed) by Sartre: "The 'essence' of Dasein is grounded in its existence" (SZ 117, MR 152). What does this mean in *Being and Time? "If the 'I' is an essential characteristic of Dasein, then it is one which must be interpreted existentially."* But then, "Dasein is, in each case, its Self only in existing" (ibid.). To put it briefly: the self is not *gegeben* (given) but *aufgegeben,* a task yet to be achieved.

THE SELF IS ENCOUNTERED IN EVERYDAY LIFE

"Everydayness" is that surface which, at the outset, was called pre-ontological, where we have a vague or "pre" understanding of the fact that in existing our Being is at stake. It is later called "thrownness" and "facticity". But looking at this daily surface, we hardly encounter ourselves as seizing upon our Being as an issue. This is what is meant by the repeated phrase *"zunächst und zumeist",* proximally and for the most part. Now, how does the self appear, proximally and for the most part? "It could be that the 'who' of everyday Dasein just is *not* the 'I myself'" (SZ 115, MR 150).

"Perhaps, in the long run, Dasein says 'I am the one', and says this loudest, when it is 'not' this being" (SZ 115).[65] The entirety of Section 27 examines the way Dasein appears to itself in daily existence. It "does" things, rather than reflecting. Daily existence is composed of running errands, performing tasks, etc. If we are essentially absorbed in our daily existence, rather than reflecting,[66] if we primarily give attention to the way we are involved in what is *no* within us, but outside in the world, then we are perhaps most of the time not ourselves. As a famous passage from *Being and Time* puts it:

> Dasein itself *is* not; its Being has been taken away by the others. Dasein's everyday possibilities of Being are for the others to dispose of as they please. The others, moreover, are not *definite* others. On the contrary, any other can represent them. What is decisive is just that inconspicuous domination by others, which has already been taken over unawares by Dasein as Being-with. One belongs to the others oneself and enhances their power. "The others" whom one thus designates to cover up the fact of one's belonging to them essentially oneself, are those who proximally and for the most part "*are there*" in everyday Being-with-one-another. The "who" is not this one, not that one, not oneself [*Man selbst*], not some people, and not the sum of them all. The "who" is a neuter, the "*they*" [*das Man*].
>
> (SZ 126, MR 164)[67]

If the self is so encountered in everyday-neutrality, then existence is somehow prescribed by the others, and not by myself. And yet, this kind of being, prescribed by the others, is what I make mine. But how? Certainly not in saying reflectively (like the character in Ionesco's *Rhinoceros*): "I want to be a rhinoceros as they all are",[68] but rather pre-reflectively, by going about one's daily business the way everyone else does, by considering the use of everyday equipment to be unremarkable in the midst of a world of microchips and intercontinental ballistic missiles.[69] Everydayness is a practical adoption of average existence.

THE SELF IS MODIFIABLE

The self, as not pre-given, and as practically determined by the "they", is modifiable. It can come back from what "they all do".

Thus, the self is the locus of the two modifications of the fabric of Dasein, which we shall soon have to examine: authenticity and inauthenticity. Authenticity should not imply that determination by "the they" will ever be overcome or eradicated. It should also not imply that solipsism or self-sufficiency will somehow come out victorious. Rather, if the triangle Self/Being-in/beings (world) is insurmountable, then average everydayness and the "they" will have to be considered as structural and not episodic determinations. At this point, what is important is simply to show how the self must be understood out of Being-in, as always already committed to its world and determined by it. Hence the three elements sketched above: the self as not given, as absorbed by the "they", and as potentially authentic.

The self and being-with *(*Mitsein*)*

There are three ways in which beings appear in the world: things that we use (equipment, *Zeug*), things that we consider ("things" here, i.e. objectively present), and others like us. "The world of Dasein frees beings which not only are quite distinct from equipment and things, but which are 'in' the world ... by way of Being-in-the-world" (SZ 118, MR 154). Thus, the "world" of Dasein is structured according to three domains: beings that are *zuhanden* (ready-to-hand), *vorhanden* (present-at-hand, or "objectively present"), and "the others". Of "the others", Heidegger says here: "[t]hese beings are neither ready-to-hand, nor present-at-hand; on the contrary, they are *like* the very Dasein which frees them, in that they are *there too, and there with it*" (ibid.). Let me begin by opposing Being-with to what is here and elsewhere called *zuhanden* and *vorhanden*.

First, we will look at Being-with (*Mitsein*) as opposed to "ready-to-hand". In Section 15, Heidegger develops what he calls "concern" with equipment. "Concern" (*besorgen*) is that aspect of care in which we find ourselves among instruments that we use. In using these instruments we precisely do not reflect on their make-up. "The kind of dealing which is closest to us is...*not* a bare perceptual cognition, but rather that kind of concern which manipulates things and puts them to use" (SZ 67, MR 95). Our everyday familiarity with the world consists in a simple engagement with clothes, pens, subways, hammers, and doors that we open and close without looking explicitly at their proportion, color, etc. Once again, this engagement is not perceptual, theoretical, or reflective, but occurs through usage.[70]

There are certainly instances where the others appear as things that are manipulated, "put to use". Any kind of master–slave relationship is of such an order. One can put "the others" to use both privately, or by the millions. In this case they become "cannon fodder". But then—and this is the point of the entire analysis of Being-with—they are not seen as *Mitdasein,* Dasein-with (title of Section 26). But even in this case, and more generally in all comportment with regard to equipment, tools, etc., the others are already somehow "there". For Heidegger, things that are simply to be put to use, *zuhanden,* belong in a context of references. The door that we open and close does not float in empty space; around doors we do not expect wheels or potato salad, but walls. And within walls, people live. This is what we mean by a "pattern of references" (*Verweisungszusammenhang*), in which others implicitly have their place. When shoes lose their soles two weeks after you bought them, you say: "They don't make things like they used to". In a thing ready-to-hand there lies "an essential assignment or reference to possible wearers, for instance, for whom it should be 'cut to the figure'. Similarly, when material is put to use, we encounter its producer or 'supplier' as one who 'serves' well or badly" (SZ 117, MR 153). In what Heidegger calls the *Zeugzusammenhang,* environmental context of equipment, the others are present in the world set up by readiness-to-hand. The world in which Dasein finds itself is always made up of things for usage, of potential others (who are not "added" to the relation of usage), and of myself. The point is that there is no "world" in the phenomenological sense, from which the others are altogether absent.

The same thing holds true of things *vorhanden* (present at hand or objectively present). We can look at the pair of shoes in such a way that we compare them with their ideal constitution, with what shoes are supposed to be. We look at them angrily when the sole has been lost: shoes are not shoes if they have no sole or heel. Thus, when a sole or heel breaks off on the sidewalk, our relation to the shoe changes, it is no longer ready-to-hand (or to foot), but "objectively present". Things "objectively present" are what science deals with. Heidegger gives the example of a hammer. As long as it works well, it is *zuhanden.* When it breaks, it becomes conspicuous, "objectively present": "If we look at things just 'theoretically', we can get along without understanding readiness-to-hand" (SZ 69, MR 98). When we look at the hammer according to its make-up—wood, iron, weight, molecular pattern, etc.—an alteration occurs, a shift in the understanding of its Being. The malfunctioning *reveals* the context of equipment. The malfunction *can* also generate the theoretical attitude. Heidegger shows this shift by starting from a statement such as "the hammer is heavy" (SZ 360, MR 412). This can mean: it is too heavy for what I want to do with it; but it can also mean that the hammer has the property of "heaviness".

> Why is it that what we are talking about—the heavy hammer—shows itself differently when our way of talking is thus modified? Not because we are keeping our distance from manipulation, nor because we are just looking away from the equipmental character of this being, but rather because we are looking *at* the ready-to-hand thing which we encounter, and looking at it "in a new way" as something present-to-hand. The *understanding of Being [Seinsverstandnis]* by which our concernful dealings with beings-within-the-world have been guided *has changed over.*
>
> (SZ 361, MR 412)

In this shift of attitude, this changing-over in the understanding of Being, lies the condition for the possibility of something like theory and, more generally, of science. It is, of course, true that other human beings can also appear in this context. The others are "present to hand" for us, objectively present, when we undertake to use them for sociological or medical research. Likewise, in our rapport with things objectively present, the others are present in a context of objective presence. A scientist is never alone in his work: Bachelard speaks of the "*cité scientifique*", Charles Peirce of the "community of interpreters" or competent speakers. In a theoretical gaze, the others are co-given. That is the point.

Yet, although in the domains of things *zuhanden* as well as those *vorhanden,* the others are present, they appear as "others" only when I look at them the way I look at myself. Being-with is discovered out of the self.

"By 'others' we do not mean everyone else but me—those over against whom the 'I' stands out. They are rather those from whom, for the most part, one does *not* distinguish oneself—those among whom one is, too" (SZ 118, MR 154). The link with everydayness is evident. In everyday life the others are "also there", there "with us". "This 'with' is something of the character of Dasein; the 'also' means a sameness of Being as circumspectively concernful Being-in-the-world. 'With' and 'also' are to be understood *existentially*" (SZ 118, MR 154–5). In the domain of Being-with, it is again the deficient mode that reveals the existential structure: "Being for, against, or without one another, passing one another by, not 'mattering' to one another—these are possible modes of solicitude. And it is precisely these last-named deficient and indifferent modes that characterize everyday, average Being-with-one-another" (SZ 121, MR 158).[71]

Here, in the context of self and Being-with, it is loneliness and solitude (which are not the same) that reveal the existential Being-with.[72] "The other can *be missing* only *in* and *for* a Being-with" (SZ 120, MR 157). Thus Being-with, as an *existentiale,* puts an end to all speculations about the constitution of other consciousnesses: "others are not proximally present-at-hand as free-floating subjects along with other things...because Dasein's Being is Being-with, its understanding of Being already implies the understanding of others" (SZ 123, MR 160f.). Only on the ground of such a structural connectedness with others, called Being-with, are phenomena such as hate, love, loneliness, as well as psychic distortions such as paranoia, etc. phenomenologically understandable.

In the context of the self and Being-with, we get the first inkling of what "authenticity" (*Eigentlichkeit*), which could also be translated as "ownness", might mean later on. "The self of everyday Dasein is the 'they-self' [*das Man-selbst*], which we distinguish from the authentic self—that is, from the Self that has been properly seized upon [*eigens ergriffen*]. As they-self, the particular Dasein has been dispersed into the 'they', and must first find itself" (SZ 129, MR 167).

Let us recall the Augustinian reference given by Heidegger concerning this entire analytic of care: the understanding of "care" "has grown upon the author in connection with his attempts to interpret the Augustinian (i.e. Helleno-Christian) anthropology with regard to the fundamental principles reached in the ontology of Aristotle" (SZ 199, MR 492). In this opposition between dispersion among the "they", and the "properly seizing upon" the self, we have the core of Augustinian anthropology. For Augustine, man is literally "dissolute", spread out among things manifold, and thus homeless, "in the region of dissimilarity". Then, through the distinction *intus-foris* (*noli foris ire*), this anthropology points the way toward "gathering oneself up" again: *intus ire* (inwardness) is what is opposed to the dispersion of the self. In going inwards, one precisely "seizes" what one most properly is. Augustine refers to this variously, e.g. *apex mentis*. By so returning upon myself, I get hold of what is most proper (*eigen*) to me.

"Authenticity" is thus a poor rendering of *"Eigentlichkeit"*. What Heidegger speaks of is the possibility of making the entire care-structure one's own, of explicitly accepting it as mine. *"Authentic Being-one's-Self* does not rest upon an exceptional condition of the subject, as detached from the 'they'; it is rather an existentiell modification of the 'they'—*of the 'they' as an essential existentiale"*

(SZ 130, MR 168). In the context of *Being and Time* (but not in that of the later *Beiträge*) it might be appropriate to speak of "ownness".

Becoming one's "own" self is even described as reminiscent of the Augustinian "turning around". *Conversio* is opposed to *aversio* (turning away from oneself) just as becoming authentic is opposed to the "they", and also as "projection" is opposed to "thrownness". "To Dasein's constitution of Being belongs *projection*—disclosive Being towards its potentiality...[It] can understand *itself* in terms of the 'world' and of others, or in terms of its ownmost potentiality-for-Being" (SZ 221, MR 264).

The self is thus not given in advance, but is rather a quest. And with quest, we are not too far from "question", the question, precisely, of Being—as the question of "its" Being. Its Being has to be made its own. The self is not a possession, but a task, not *gegeben,* but *aufgegeben.* It is always ahead of us as a possibility; it is always forthcoming, always in advent—like time. Thus, a peculiar link appears at the end of this analysis of care, a link between the interrogative character of the self, the question of Being, and the futurity of time. All three point towards a not-yet. I am what I ask about.[73] But what I ask about, what is always ahead of me, and what—once again—generates a particular wholeness of our Being, is death.

Temporality

The first general title for the structural whole of Dasein was Being-in-the-world. Because this was "too formal" it was not replaced, but deepened by a second title, care. Care "filled in" the inner complexities of the transcending structure, the existing structure of Dasein. It diversified Being-in-the-world according to the many *existentialia,* only a few of which I have developed ("understanding", in particular, will have to be taken up again in a later context). In particular, "care" indicated something that Being-in-the-world did not; namely, that Dasein is always ahead of itself (*sich-vorweg*). Correlatively, this means that Dasein is never "complete" (*abgeschlossen*). "As long as Dasein *is* as a being, it never reaches its 'wholeness'. But when it gains such wholeness, this gain becomes the utter loss of Being-in-the-world. In such a case, it can never again be experienced as a being" (SZ 236, MR 280). Once again, a deficient mode is revelatory of an existential structure. Temporality is revealed by the utter annihilation of Dasein as a being, i.e. by death.

The heuristic function of death

If care is the general name for our Being-ahead of ourselves, a way of making this concept more explicit would be to look more closely at death.[74] In the same way as "care" was said to concretize "Being-in-the-world", we can expect a concretiz-ing of "care" from an existential analysis of death. Dasein concretely adopts a comportment with regard to the not-yet: "The uttermost 'not-yet' has the charac-ter of something *towards which* Dasein *comports itself* " (SZ 250, MR 293).

Death, perhaps more than anything else, can be dealt with as if it were an object, even a remote one. This is the reason why funeral "celebrations" are so important. They prove to us that we are very much alive, and that death is something that only happens to others. The "existential" analysis, cannot be content with such a reification of death. "Death is not something not yet objectively present, nor is it that which is ultimately still outstanding" (SZ 250, MR 293). If we look at comportments such as funeral "celebrations", or the way we treat terminal patients (locking them away in specialized sections of hospitals), or the way we treat our own death in relation merely to dates (asking when it will come)—we see, of course, that there is a very definite comportment with regard to death, and that we deal with it constantly. Death may even have such a presence that without it we would not run around keeping track of time, and talk about "losing time", as we do. All this is much developed, although in a different vein, in Simone de Beauvoir's *All Men are Mortal*.[75] She describes a man who is exempt from dying. He is carried around by others, does not act, and is utterly amorphous, the moral of the story being that the imminence of death is what stimulates action. In Heidegger's more structural interpretation death is something "impending" (*ein Bevorstand*). "For instance, a storm, the remodeling of the house, or the arrival of a friend, may be impending; and these are beings that are respectively present-at-hand, ready-to-hand, and there-with-us. [However] the death which impends does not have this kind of Being" (SZ 250, MR 294). Rather, death is Dasein's "ownmost potentiality". "Death is a possibility-for-Being that Dasein itself has to take over in every case" (ibid.).

What is at stake here is the fore-structure of Dasein. *Vorgriff, Vorsicht, Vorhabe* have revealed that Dasein is always ahead of itself. Thus, what is necessary is to "set forth the *formal* structure of end in general and of totality in general" (SZ 241, MR 285). Death answers these two questions: how is Dasein ultimately ahead of itself (towards its end) and how can Dasein ultimately be whole? It should be clear that these two questions arise from the analyses of both Being-in-the-world and care. It is in these two contexts that the not-yet and the potentiality for Being-whole appeared as existential determinants.

What does it mean then, to understand death existentially—i.e. in the way that other structures have been understood? Certainly not as merely something to come. The not-yet indicates a reference to the present moment. It is now that I am "not yet" dead. But that not-yet is no insignificant matter. On the contrary, in this impending character of death, it *is* present; present as my ownmost possibility. Hence, death is not understood existentially if I understand it simply as the ending of Dasein, some day, some place, due to some cause. "The 'ending' which we have in view when we speak of death does not signify Dasein's Being-at-an-end, but a Being-towards-the-end" (SZ 245, MR 289). Thus death is a structural determinant of Dasein wherever and however it exists. Further, "Death is a way to be, which Dasein takes over as soon as it is" (SZ 245, MR 289).[76]

The heuristic function of Being-towards-the-end is to reveal Dasein's temporality. The way temporality is discovered in the Analytic is noteworthy. It is neither through memory and its retrieval of the past nor through the observation

of movements and their number according to a before and an after. It is the imminence of death that is operative in this discovery of time and not the schema of
Aristotle's *Physics,* nor that of Augustine's *Confessions.* It is significant that if
there is any recognition of birth as an *existentiale* in *Being and Time,* it is due to
Being-towards-death, and not the other way around (SZ 374, 390–1). Birth would
be the facticity of a tradition in which we stand, the heredity of family and country
into which we are born, a communal and historical heritage that roots us, whatever
we may do to eradicate them. Since the temporality of Being is the climax of the
fundamental ontology, the analysis of death is preparatory in a central way. It prepares the temporality of Dasein, which in turn was to make thinkable (although
this is not carried out in *Being and Time*) the temporality of Being itself.

Heidegger characterizes the main traits of Being-towards-the-end by saying:

> With death, Dasein stands before itself in its ownmost potentiality-for-
> Being. This is a possibility in which the issue is nothing less than Dasein's
> Being-in-the-world...When Dasein stands before itself as such a possibil-
> ity, it has been *fully* assigned to its ownmost potentiality for Being. When
> it stands before itself in this way, all its relations to any other Dasein have
> been undone. This ownmost, non-relational, possibility is at the same time
> the uttermost one. As potentiality-for-Being, Dasein cannot outstrip the
> possibility of death. Death is the possibility of the absolute impossibility
> of Dasein. Thus, death reveals itself as *that possibility which is one's own-
> most, which is non-relational, and which is not to be outstripped.*
>
> (SZ 250–1, MR 294)[77]

Death is my ownmost possibility, that is to say, it is always mine. In death, at
least, I cannot do as everyone else, as "they" do. One cannot die as "they all"
die. Thus, death is also a non-relational possibility in Being-towards-the-end. In
death, I am not related to others, I am always thrown back upon myself. Death is
the initial breach from the everyday realm of common sense. It is a transcending movement, but not towards the everyday world; rather it is, as my ownmost
possibility, a transcendence that pulls me towards my own self, it is a transcending movement of self-appropriation, of self-authentication. It "cannot be
outstripped", i.e. there is nothing more encompassing than this Being-towards-
the-end. It "owns up" to the whole of existence from birth to death. These three
determinations of Being-towards-the-end, as ownmost, as non-relational, and as
unüberholbare possibility, show the importance of death for any understanding
of Being. "Higher than actuality stands possibility" (SZ 38, MR 63). But the
highest possibility is always ahead of us: the possibility of becoming total in the
form of total negation, death.

The heuristic, or revelatory function of Being-towards-the-end, demonstrates,
in a first approximation, that death reveals our temporality; and does so in such a
way that the main trait of time is what is not yet, what lies ahead, the future. The
future is not the coming of various events, but it is Dasein itself that is "to come",

that is imminent. "If Being-towards-death...belongs to the Being of Dasein, then this is possible only as 'to come'" (SZ 325). This futurity is the basic trait of projection. "Self-projection upon the 'for sake of oneself' is grounded in the future, and is an essential characteristic of existentiality. *The primary meaning of existentiality is the future*" (SZ 327, MR 375–6).[78]

But death is not only indicative of Dasein's fore-structure, of futurity—it is not only heuristic for the temporality of Dasein.[79] In this sense, death is more central to the analysis than dread. Death has also to do with Dasein's possible totality or wholeness.

The totalizing function of Being-towards-death

The three traits of the possibility of death just mentioned—ownmost, non-relational, and never to be outstripped—indicate that Being-towards-death is not merely one possibility among others. Likewise, death is not just one horizon of projection among others. Death is that horizon of projection within which all projects are made possible. Any possibility that can positively be seized by Dasein has Being-towards-death as its ground. That is what is meant by "never to be outstripped", *unüberholbar,* not surpassable. This possibility cannot be surpassed because it itself surpasses all other possibilities. It permeates the totality of each and every possibility of understanding Being.

Death is, in this sense, a possibility of understanding. In Being-towards-death we have always already understood that our Being is at stake in whatever we undertake. Also, we have always already understood that we can be whole, but only in the form of total negation. Heidegger exposes this by a new utilization of the structure of dread. "Dread in the face of death is dread in the face of that potentiality-for-Being which is one's ownmost, non-relational, and not to be outstripped. That in the face of which one has dread is Being-in-the-world itself; this dread...amounts to the disclosedness of the fact that Dasein exists as thrown Being *towards* its end" (SZ 251, MR 295). That *of* which one has this dread is simply Dasein's potentiality for Being.

To say that "that of which one has this dread is Dasein's potentiality for Being" (ibid.) is to say that Dasein already "understands", albeit implicitly, that Dasein can also not be. The Being-ahead of oneself thus appears as Being-towards-death: "This element in the structure of care has its most originary concretion in Being-towards-death" (ibid.).

To be no longer ahead of oneself would be to have become "whole". Futurity is that lacuna within Dasein that makes it incomplete. Since there is always something outstanding, it is never entirely itself. But there is a moment when there is no longer anything outstanding, when futurity is abolished and the lacuna is filled. This is the moment of death. This is a "moment" in the double sense of a particular instant in duration, as well as of an "aspect", a structural element. Being-towards-death is that moment coming upon Dasein in which the annihilation of futurity is possible, is lived as a potentiality.

These remarks indicate that we must not take the *Ganzheit* of Dasein as the sum total of the events of a lifetime. To say that Being-toward-death totalizes Dasein is not to say that only in death is a life a whole. Thus Heidegger, in his continuous attempt to reach a more originary level, asks: "Have we brought the 'whole' of Dasein into the fore-having of our analysis? ... It may be that as regards 'Being-towards-death' the question may have found its answer" (SZ 372). It "may be", but in fact is not the case if we understand death merely as the "other end" of the life span, opposed to birth. The in-between birth and death, the extension and completion in ending one's life, is not the way death totalizes Dasein.

Death totalizes Dasein as its possibility. "Dasein does not exist as the sum of momentary actualities of experiences which come along successively and disappear...Factical Dasein exists as born; and, as born, it is already dying, in the sense of Being-towards-death" (SZ 374, MR 426). The possibility of totalization is an *existentiale*, it is a structural determination. Although it is not the case that such a possibility can be actualized,[80] it is still true that this *Ganzseinkönnen* can be or not be authentic, i.e. made mine as possibility, or not. Although factically I am already "stretched out", to make this possibility explicitly mine is to "stretch oneself out". Dasein *is* its possibility for totality when it is not merely stretched along, but when it stretches itself along. Such being-stretched and stretching-itself is what Heidegger calls *Geschehen*.[81] "The specific movement in which Dasein *is stretched along and stretches itself along,* we call its '*happening*'" (SZ 375, MR 294). The possibility of *Geschichte,* history, is grounded in such *Geschehen.*

Such extending-extendedness is called the "most originary concretion" of care (SZ 251, MR 294). Indeed, *Being and Time* does not step further in the quest for an originary structure of Dasein than the whole that Being-towards-death reveals, the whole of our Being as temporality.

Temporality as the "sense" of care: the three "ecstasies"

Care was the structure of Dasein that appeared as more "articulate" than the structure called Being-in-the-world. Indeed, its inner articulation is so complex that it is very difficult to map out the exact configuration of the various *existentialia.* Heidegger said of this articulation: "The ontological question must be pursued still further back until the unity of the totality of this structural manifoldness has been laid bare. The originary unity of the structure of care lies in temporality" (SZ 327, MR 375). Or, as the title of Section 65 puts it: "Temporality as the Ontological Meaning [*Sinn*] of Care" (SZ 323).

The full regulatory constellation of determinations is then given with the notion of temporality. Temporality is only the extreme form of Dasein's extendedness, which has been the chief subject matter ever since the first page of the Analytic where it is said that "*the 'essence' of Dasein lies in its existence*" (SZ 42, MR 67).

103

Temporality thus retrieves the previous analyses. The entire second division is intended to show that care is temporal—"that all fundamental structures which we have hitherto exhibited, are at bottom temporal" (SZ 304). Of this retrieval, by which Division Two repeats Division One, several elements are noteworthy.

PROJECT

Projection is possible only on the grounds of futurity. To be ahead of oneself "is possible only in that Dasein *can* come towards itself in its ownmost possibility," i.e. in Being-towards-death. "This letting-itself-*come-towards*-itself is the origi-nary phenomenon of the *future*...By the term 'futural' we do not here have in view a 'now' which has *not yet* become 'actual' and which some time *will be* for the first time; rather we have in view the coming [Kunft] in which Dasein, in its own-most potentiality for Being, comes towards itself" (SZ 325, MR 372–3).[82]

Thus, futurity lies at the bottom of ek-sistence, care, Being-in-the-world, trans-cending, pro-jection. Futurity is the condition of the possibility of projection. From this, in the Analytic, it becomes apparent that the future is the guiding fig-ure of temporality. Projection had the meaning of finding myself—of being an issue for myself. Now, "Self-projection upon the 'for-the-sake-of-oneself' is grounded in the future. It is an essential characteristic of *existentiality. The pri-mary sense of existentiality is the future*" (SZ 327, MR 375f). The future is that figure of time that is connected to possibilities. And, Dasein exists only as possi-bility. "Higher than actuality stands possibility", or potentiality (SZ 38, MR 63). Here, in the context of temporality, "Self-understanding that projects itself in its existentiell possibility can be such only on the ground of the future" (SZ 336).

THROWNNESS

Just as projection appears to be a structure made possible by an aspect of tempo-rality, so too is thrownness. We are thrown into the world—i.e. always already there, carrying our past with us. "Taking over thrownness is possible only in such a way that Dasein can be as it already was, making this its own" (SZ 326). Thrownness and facticity now appear as derived from a form of time, the determi-nation by which Dasein *already was* this or that. To take over thrownness is to say "yes" to such having-been; to make such having-been present. Put more strongly, it is to make having-been the possibility towards which Dasein projects itself. "Only insofar as Dasein *is* as an 'I-am-as-having-been' can Dasein come towards itself futurally" (SZ 326, MR 376). Past and future are thus intertwined in a peculiar manner. Past is possible only on the basis of futurity; it is somehow derived from the latter since what I was stands before me as yet to be taken over by me. This stands against a certain tradition that was emerging at the time Heidegger wrote *Being and Time*. The psychic (or social) past, is not seen by Heidegger as deter-mining; rather, the past is a set of possibilities ahead of Dasein. That is to say: these possibilities are to be made our own. Psychic or social determinations are not viewed as an inescapable fate as in the case where someone born into a class remains bound to that class, or in the case where a psychic structure is formed at an early age and remains with one, haunting one like a ghost.

"'As long as' Dasein factically exists, it is never past [vergangen], but it always is indeed as already having *been* [gewesen]" (SZ 328, MR 376)—which is to say that what has been, still "essences" (*west*), is still actively there as a possibility. Thus, by "*Vergangen*" we understand that only things present-at-hand can pass away, whereas by "*Gewesen*" we understand a continuing process.[83] "The primary existential meaning of facticity lies in 'having been'" (SZ 328, MR 376). The past is not a dead weight that I carry around, but rather is alive as my potential.

BEING-WITH OR ALONGSIDE

The care structure revealed Dasein as present *to* instruments, tools, objects, and other people. They are co-present. Being-with or alongside other such beings is again possible only on the basis of a particular temporal structure, what we usually call the "present". "Being-alongside what is ready to hand...is possible only by such a being *present* [in einem Gegenwärtigen]" (SZ 326, MR 374). The point is that the present, too, contains a process, which is poorly translated as "making" present.[84] This process once again reveals the primacy of the future, since the present is such only on the ground of potentiality, of letting-things come towards me. To be present to things around us, then, is to let ourselves be "encountered undisguisedly" (SZ 326, MR 374).

Future, past, and present thus appear as the three fundamental modes of *Erschlossenheit* (disclosedness). When they are taken over explicitly—i.e. made my own—they are the modes of *Entschlossenheit* (resoluteness). Because of the primacy of the future, we shall have to see that the most appropriate title for authenticity is "anticipatory resoluteness". This concept has sometimes been misinterpreted[85] as if resoluteness or resolve meant the decidedness of going ahead, despite everything and everyone. If one misses the relation between disclosedness and resoluteness, one will classify Heidegger as a "decisionist". But resoluteness is only a name for a certain modification of temporality—authentic temporality. The concept of temporality now synthesizes the three figures of time just drawn out: "Coming back to itself futurally, resoluteness brings itself into the situation by making present. The character of 'having-been' arises from the future ... This phenomenon has the unity of a future which makes present in the process of having been; we designate it as '*temporality*'" (SZ 326, MR 374).

The concept of temporality thus explicitly indicates the threefold standing-out of oneself into future, past, and present, which are at the root of projection, thrownness, and Being-with.

We now see more clearly the sense in which the preliminary idea of the Analytic was appropriate when it was labeled "existentiality". Indeed, Being-in-the-world is to stand out of oneself. Literally, to stand out means ek-stare. Future, past and present are ek-static structures of the self. By "future" is meant Being-towards-oneself.

Table 2.5 The ek-static structures of the self

Future	Projection	Being-towards-oneself
Having-been	Thrownness, facticity	"Back-to"
Present	Being-with	"Letting oneself be encountered by"

By "having-been" is meant back-to. By "present" is meant letting-oneself-be-encountered-by. Thus, we can represent the relations as shown in Table 2.5.

"The future, the character of having been, and the Present, show the phenomenal characteristics of the 'towards-oneself', the 'back-to', and the 'letting-oneself-be-encountered-*by*'. [These three phenomena] make temporality manifest as the *ekstatikon* pure and simple" (SZ 328, MR 377). Heidegger summarizes: "Temporality is the primordial 'out-of-oneself' (in and for itself). We call the phenomena of the future, of having-been, and of present, the *ecstases* of temporality" (SZ 328, MR 377).

The thrust of this entire presentation modifies the transcendental approach. It indicates how the content of the care-structure is possible. The "possibilizing" ground is not discovered in the subject, but in man's connectedness with the world. It is due to the radical finitude that Heidegger opposes to the nineteenth-century tradition, that this connectedness, this contextuality, should be formulated in terms of time. If man is entirely temporal, he is entirely finite. It is a consequence of Heidegger's transformation of the philosophy of subjectivity that the most originary origin is located *in forms of temporality*. To say that temporality is "the most originary concretion" of care, is to say that man is thoroughly finite.

The concept of ecstatic temporality is perhaps what is most innovative in *Being and Time*. One has to measure it against the inherited notions of time. These are mainly organized around two models: the physicalist model and the mental model. The physicalist model says that "time is the number of movements according to before and after".[86] The mental model says that time is *distentio animi*. As Augustine writes, "It seemed to me that time is nothing else than extendedness; but of what sort of thing it is an extendedness, I do not know; and it would be surprising if it were not an extendedness of the mind itself". Augustine adds: "Time is threefold, namely the present of things past, the present of things present, and the present of things to come".[87] Time is the extendedness of the mind because the mind can have things past, present, and yet to come present before its inner eye. But in both models, *praesto habere,* the present, is the guiding mode of time.

Heidegger breaks with the linear representation of time *because* he breaks with the pre-eminence of the present. When time is primarily understood out of what is given, presently there, future and past appear as mere extensions of that givenness. This is true until Husserl, who in that respect belongs very much to the Augustinian tradition. Husserl's treatment of time is more metaphorical than conceptual. Most of it follows directly from his notion of *Erlebnisstrom*.[88] The "stream" and "flux" of time—linear time—are extensions of the original presence

(*Urpräsenz*). The now-moment is the point of intersection between past moments and future moments, between retentions and protentions. This is typical for the entire metaphysical tradition. The pre-eminence of the present, a linear representation of flux, and the conception of past and future as extensions of the present, belong to one another and form a system. In Kant, at least—as opposed to the idealist Husserl—there was an effort to reconcile the Aristotelian and the Augustinian traditions (the physicist and the mentalist model) through the concept of *Beharrlichkeit* (permanence).[89] Husserl, in focusing exclusively on inner time consciousness, actually reverts to Augustine.

Heidegger transforms the linear representation of time into an ecstatic one, emphasizing the pre-eminence of the future over that of the present; and instead of seeing past and future as extensions of the present, they now appear as three co-originary, "equi-primordial" moments (SZ 329, MR 378).

Ecstatic temporality as the condition for history

Temporality is not a "thing" that happens at some "time" to something objectively present. "Temporality is not, prior to this [the threefold ecstasis], a being that first steps out of itself; rather its essence is a process of temporalizing in the unity of the ecstases" (SZ 329, MR 377). Thus, the "existential" directionality (the three directions) is originative with regard to common time as a succession of moments. This common time is called here "vulgar" time, just as the common sense of "phenomenon", at the beginning of the book, was called "vulgar". Common time "levels out" this originative time. Ecstatic temporality is originative because it makes past, present, and future possible. When these are leveled out, or considered in isolation, we have common time. Ecstatic temporality lets time emerge (*entspringen*); ecstatic temporality is therefore originary time (*ursprüngliche Zeit*).

Time is not the sum total of the three ecstasies, rather "*in every ecstasis, temporality temporalizes itself as a whole*" (SZ 350, MR 401). That is to say, there is no *pure* past, present, or future. All three are inextricably interwoven. Heidegger can therefore claim that temporality, in any of its ecstases, lays the ground for the care structure. Any ecstasis contains the whole of ecstatic temporality. "*In the existential unity with which temporality has fully temporalized itself, in each case, is grounded the totality of the structural whole of existence, facticity, and falling—that is, the unity of the care-structure*" (SZ 350, MR 401). The transcendental condition of possibility of facticity is "having-been", that of falling is "present", and that of existence is "futurity". But just as each of these three determinants of care is "held" by the two others, so is each temporal ecstasis "held" by the two others (SZ 350, MR 401).

When time is taken as a "whole" in the threefold ecstasis, originary time is most effectively distinguished from our usual understanding of time as "pure sequence, without beginning or end, of now moments" (SZ 329). This shift exhibits the "understanding of Being" in a new light. Insofar as I understand myself as having-to-be, i.e. as always ahead of myself, the future arises. But this future *is* my understanding of Being, since in such futurity I know that it is my Being that is at stake or is at issue. In the determinateness of that Being which is at issue—determinateness by heritage, birth, culture—arises the past, the "having-been". Thus, the future comes back to the past, to that whole that my understanding of Being has already been. Finally, in turning towards individual things that are co-given with me, the present arises. But here again, this ecstasis is but an aspect of the understanding of Being, where I am for myself an issue in so far as I am "there", together with things and others.

This shows that *Seinsverständnis* has been unfolded as originary time. If, at the beginning, the "guiding idea" of Dasein was "existence", i.e. the Being of Dasein, it is now clear why "existence" designates a radical openness, standing-out of oneself, which has now been specified as the unity of the three temporal ecstases. If Dasein is constituted at its core by such a threefold temporal reaching-out, then we also understand the word "Dasein" better. "Ecstatic temporality 'clears' [lichtet] the 'there' originarily" (SZ 351, MR 402). Ecstatic temporality assumes the function that transcendental apperception (or according to *Kant and the Problem of Metaphysics,* imagination) assumed for Kant. We have interrogated, Heidegger says, "the whole constitution of Dasein, that is, care, according to the unitary ground of its existential possibility...[Ecstatic temporality] is what primarily regulates the possible unity of all Dasein's existential structures" (SZ 351).[90] Just as transcendental apperception is the unity of all functions of the knowing subject in Kant, ecstatic temporality is the unity of all functions—or structures—in *Being and Time.* But of course, ecstatic temporality has a completely contrary meaning to transcendental apperception, since Kant's notion of apperception is both worldless and timeless. Kantian language, however, abounds in this context in *Being and Time.* "This temporality is the a priori [im vorhinein] condition of possibility of Being-in-the-world" (SZ 351, MR 402).

Existence indicates the temporal sense—the temporal, threefold directedness—of our understanding of our Being. It is thus clear why "existence" is, from the beginning of *Being and Time,* that idea that can lead to Dasein's genuine understanding of itself.

As I have indicated—though we cannot pursue this at length—the ecstatic constitution of Dasein is the basis for history. Beyond the play on words—existing or ecstasis (*Geschehen*); history (*Geschichte*)—it is clear that transcendental apperception in the Kantian sense can never "historicize" itself. On the contrary, since the core of Dasein has been revealed to be worldly and temporal, this core is ultimately what makes history possible.

Furthermore, this core, which replaces transcendental apperception, is what makes the understanding of Being as an *event* possible. Rather than speaking of Being as a noun, one will have to speak of it as a verb. Perhaps one should translate the title *Sein und Zeit* as "To-be and Time". At any rate, *Seinsverständnis* is an event, a *Geschehen.*

The understanding of Being, as tied to others with whom I live—among whom I am "thrown"—is an event of Being-with. Since I find myself always already among fellow men, they further concretize my facticity. Heidegger calls this temporal dimension of Being-with others "fate" (*Schicksal*). "If fateful Dasein, as Being-in-the-world, exists essentially in Being-with others, its historicizing [Geschehen] is a co-historicizing [Mitgeschehen]; and it is determined as destiny [Geschick]. This is how we designate the event of a community, of a people" (SZ 384, MR 436). To exist historically is to exist as a community. This further reveals the limiting sense of thrownness. I am thrown into a community, I am limited in my autonomy, limited to the possibilities that "destiny"—understood as such a principle of limitation—prescribes to me.

The "ontological understanding of historicity" (SZ 375, MR 427) thus depends entirely on the understanding of temporality as ecstatic. Projection is historical in so far as it "comes back" to what has been, to what Heidegger calls our "heritage" (*Erbe*). This concept stands for thrownness in historical analysis. "Theories of history", or "philosophies of history", are but thematic developments of ecstatic temporality. To live explicitly as historical—to have historical consciousness—is to explicitly assume, or deal with, one's heritage and to gather from it one's possibilities for projection. Such explicit historicizing is called "retrieval" ("*Wiederholung*") "a return into the possibilities of a Dasein that has been" (SZ 385). This is not a view of history that is oriented backwards, since to "retrieve" the past consists in actually "responding" (*erwidern*) to it. "The retrieval responds to the possibility of that existence which has been there" (SZ 386). "In the retrieval, fateful destiny can be disclosed explicitly as bound up with the heritage that has come down to us. It is only by this retrieval that Dasein can make its own history manifest. This happening is, as such, grounded existentially in the fact that Dasein, as temporal, is open ecstatically" (SZ 386).

With this core of historicizing, ecstatic temporality, the transcendental analytic of Dasein has reached its deepest level. On three levels—Being-in-the-world, care, and temporality—the understanding of Being has been uncovered with increasing concreteness. The most concrete way in which Being is always an issue for Dasein lies in its futurity.[91] The many formal elements—called *existentialia*—by which *Being and Time* characterizes our Being have been tied back to one simple knot: ecstatic temporality. It is from this core of Dasein that we also have to understand the ontic modifications that have already appeared in the analysis, namely authenticity and inauthenticity.

The ontic modifications of the understanding of Being

Let us begin the analysis of authenticity and inauthenticity with a remark that may seem out of place. The three chief *existentialia* that structure "Being-in" are attunement, understanding, and speech (SZ 133), and later, attunement, understanding, and falling (SZ 346). The change seems to be due to the transition from Division One to Division Two. Indeed, in Division Two, we have the equations:

"Just as understanding is made possible primarily by the future, and attunement by having-been, so...is falling made possible by the present" (SZ 346). This is, in fact, only one of the many difficulties that a "systematic" reading of *Being and Time* encounters. It may well be that *Being and Time* should not be read that way. Nonetheless, there seems to be a good reason for the shift in question. Indeed, Heidegger addresses himself to this in the Section entitled "The Temporality of Speech" (Section 68d). Prior to this section he dealt with the temporality of understanding, attunement, and falling—those *existentialia* that, in the context of the second division, are presented as primary. In the section on the temporality of speech, we read this astonishing sentence: "the temporality of speech, that is, of Dasein in general" (SZ 349, MR 401). It appears, then, that speech now stands for the temporal constitution of Dasein altogether.[92] This is so because speech is not only indicative of Dasein's temporality, echoing, as it were, what we always live ecstatically. Speech seems to be, at least here, another name for Dasein. Perhaps this prefigures developments in Heidegger's later writings. The difficulty, at any rate, is noteworthy. He writes: "We can define the ontological meaning of the 'is', which a superficial theory of propositions and judgments has deformed to a mere 'copula'. Only in terms of the temporality of speech—that is, of Dasein in general—can we clarify the genesis of 'signification'; and only thus can we clarify ontologically the possibility of concept-formation" (SZ 349, MR 400–1).

We are not interested here—and in fact Heidegger does not pursue this in *Being and Time*—in the temporality of speech *as* the temporality of Dasein. I merely want to give a plausible interpretation of the shift from one triad (attunement, understanding, speech) to another triad (attunement, understanding, falling). This interpretation would maintain that speech designates the temporal constitution as such, and that "falling" accordingly stands for the care-structure that is rooted in the "present", which is one of the three ecstases.

In this section on the temporality of speech, Heidegger suggests a chart that incorporates the three basic traits of care—but transformed into understanding/ attunement/falling—the three ecstases in which they are rooted, and the ontic modifications, authentic or inauthentic, corresponding to each. This chart is reproduced here as Table 2.6.

One must ask how this distinction between two "modifications" of Dasein arises at all.[93] What makes it necessary? We know that the modifications are of *Seinsverständnis*. This results from the way the book begins. Methodologically, Heidegger examines everyday forms of life—of "Being there"—and then

Table 2.6 Heidegger's three basic traits of care, ecstases and the ontic modifications

Structure of care	Primary ecstasis	Inauthentic mode	Authentic mode
• Understanding	• Future	• Awaiting	• Anticipation
• Attunement	• Having-been	• Forgetting	• Retrieve
• Falling	• Present	• Instant	• Making-present

shows how these structures can be made explicit. The distinction between implicit understanding (or pre-understanding) and explicit understanding (or "knowledge", as he says at the beginning of *Being and Time*) indicates that our understanding of Being can be modified. Such was precisely the subject matter of Section 4: "The Ontical Priority of the Question of Being" (SZ 11, MR 32). As Heidegger says there, "Only when the inquiry is itself seized upon in an existentiell manner as a possibility of Being of each existing Dasein, does it all become possible to disclose the existentiality of existence" (SZ 13, MR 34). But to explicitly take up the question of Being—that is, to practice fundamental ontology—can hardly be the sole modification of *Seinsverständnis;* rather, in accordance with the general project of the Analytic, such explicit inquiry must, in its possibility, be rooted in a general disposition of Dasein. This general disposition can be developed on the basis of two "modes" that we have already encountered: *Ganzseinkönnen* and *Entwurf.*

One must see that these two "modes of Being", as they are also called, necessarily arise once Dasein is equated with its possibility, or its potentiality; once "possibility" is no longer opposed to "actuality". As Heidegger says, "That being for which its Being is always an issue, comports itself to its Being as towards its ownmost possibility" (SZ 42), i.e. "Dasein *is* its possibility...And because Dasein is in each case its own possibility, it *can* choose itself and win itself; it can also lose itself and never win itself" (SZ 42, MR 68). It is here that we come across the first mention of authenticity and inauthenticity: "[O]nly in so far as it is essentially something which can be authentic—that is, something of its own—can it have lost itself" (SZ 43, MR 68).

The task now is to consider the formal fundamental structure in so far as it can be "colored" in one way or the other: tracing back the *Entsprungene* to the *Ursprung.*

Totality and structure

In the section that drafts the general outline of the Analytic of Dasein, there is a first mention—although implicit—of Dasein's possible inauthenticity: "Dasein has the tendency to understand its own Being in terms of that other being out of which it comports itself proximally...in terms of the 'world'" (SZ 15). This means, as Heidegger explains later, that Dasein does not understand itself as a whole, "proximally and for the most part", but understands itself through the particular things it is involved with. Heidegger calls this phenomenon of being lost in the world "absorption in the world", *Aufgehe.* This will have to be opposed to the self-understanding Dasein achieves when it takes the world as a whole. This "whole" world is not the world referred to in the above quotation—the latter is not to be understood in the existential sense, since in being absorbed in this way Dasein and the world do not comprise a "whole".

Inauthentic totalization[94]

The absorption in things confronting us is precisely the inauthentic mode of Being-in-the-world. Inauthentic Dasein understands itself in terms of "an objective context of things" (SZ 201). Thus, the inauthentic or "improper" modification of Dasein is such that Dasein understands itself as if it were something objectively present among other things. But in such absorption, we still have one specific modification of the "*existentiale*" called "wholeness". Heidegger describes inauthentic wholeness by saying, "Dasein *loses itself in such a manner that it must, as it were, only subsequently pull itself together out of its dispersal*" (SZ 390, MR 442). That is, Dasein is a whole by summation, summarizing objectively-given experiences. One sits back and wonders about the idiocies one believed in years ago, sums up past events, and then concludes: this is all me. Heidegger writes: "one thinks up a unity in which that 'together' (that whole) can be embraced" (ibid.).[95]

In inauthentic totality, one does not see the forest for the trees. Later in his career, Heidegger will call this inauthenticity—when "Dasein" has become "Thinking"—"calculative thinking".[96] The totality of Dasein, in inauthentic existence, is the mere result of what Heidegger ironically calls "exact thinking". Such totalization of Dasein proceeds by anecdotal cumulation; but such calculus, "adding" fact to fact, can never reveal Dasein as a whole. Only things *vorhanden* can be so compiled. What is "exact" about this type of approach is its splitting of the whole of Dasein into singular, "objectively given" data.

How precisely does inauthentic existence overlook totality? Two points are salient here. First, bits and pieces of a biography can appear as such only because Dasein is always already anticipatory. It is an act of "whole" anticipation that Dasein, as "calculative", has transformed all that there is into *Vorhandenes*. Inauthentic "wholeness" thus misses the totalizing character of its own project, of its own understanding of the world as a sum total of data. Second, inauthentic existence misses the fact that one can only speak of the dispersion of that which was once whole. The cumulative constitution of Dasein's wholeness is possible only because Dasein is structurally already a whole. Our everyday lives can become atomized only insofar as we are originarily a whole. Only a being that is originarily a whole can subsequently be put together "objectively" as though it were a puzzle.

Inauthentic totality appears most clearly from the point of view of temporality. It has been said that the present is that determination of time that is primary in things, "present at hand", objectively given, *Vorhanden*. The atomization of the self into verifiable fragments only results from an understanding of time as atomized, as composed of datable moments. Heidegger understands the process of "dating" or "datability" (SZ 407, JS 374), inauthentic existence, and the primary ecstasis of the present, as parallel.

The composing elements of existence are not a "structure", although the whole of Dasein constituted a posteriori in a calculative manner is an empirical whole. The originary contextuality of future, past, and present retreats when we investigate "what is the case". "Dating relates to something present-at-hand" (SZ 416–7). "The now-moments are somehow co-given [with Dasein]: i.e. beings are encountered, and so are the now-moments" (SZ 423). Biographical summation is only the consequence of an understanding of time as linear, as constituted by a sum of now-moments.

Inauthentic temporality thus appears as "a sequence of 'nows' which are constantly 'present-at-hand', simultaneously passing away and coming along. Time is understood as a *succession,* as a *flowing stream* of *nows,* as the 'course of time'" (SZ 422, MR 474). With such an understanding of Being—*qua* understanding of time—Dasein covers up the totalizing function of the present. It forgets that time is a whole only in the threefold ecstasis. Inauthentic temporality is the representation of a time endlessly running, of time as "*a free-floating-in-itself of a course of 'nows' which is present-at-hand*" (SZ 424, MR 476).

Authentic totalization[97]

In the authentic modification of the existential fabric, Dasein's wholeness is not reached through a compilation of data. The singular possibilities in which Dasein engages stand in a structural relationship with one another. Authentic existence and authentic temporality are nothing but the explicit meaning of that structural whole. Dasein does not affix itself to one or another particular possibility. Rather, all its concrete possibilities are unified out of Being-towards-death, out of the future. "Can Dasein also exist *authentically* as a whole?" (SZ 234, MR 277).

The condition for such unification is what Heidegger calls "'anticipation' of the possibility [of death]" (SZ 262, MR 306). Such anticipation locates, so to speak, any particular undertaking within a totality that arises from the future.

> When, by anticipation, one becomes free *for* one's own death, one is liberated from one's lostness in those possibilities which may accidentally thrust themselves upon one; and one is liberated in such a way that for the first time one can authentically understand and choose among the factical possibilities lying ahead...Anticipation discloses to existence that its uttermost possibility lies in giving itself up, and thus it shatters all one's tenacity to whatever existence one has reached...Since anticipation of the possibility which is not to be outstripped discloses also all the possibilities which lie ahead,...this anticipation includes the possibility of taking the *whole* of Dasein in advance in an existentiell manner; that is to say, it includes the possibility of existing as a *whole potentiality-for-Being* [ganzes Seinkönnen].
>
> (SZ 264, MR 308–9)

As opposed to "calculative" totalization, this is the model for the later concept of "releasement" (*Gelassenheit*).

The distinction Heidegger uses is that between *Summe* and *Gefüge:* two modes of totalizing, one inauthentic, the other authentic. The concrete possibilities are seen in their reciprocal relation. The context within which Dasein lives is such that it is able to tie together all the possibilities that stand open for it. That universal horizon is the possibility of death. The constitution of the totality of Dasein occurs in the anticipation of death. All concrete possibilities stand in relation to that ultimate possibility. Authentic existence is, in the same instance, Being-towards any given concrete possibility and Being-towards death.

Just as dread revealed the formal totality of care and Being-towards death the formal totality of temporality. Here too, one peculiar existential determination reveals the possibility of authenticity: the call of conscience (Section 56).[98] Conscience is an originary phenomenon of Dasein because it ties every possibility back to its temporal "origin", which is the ecstases in their unity. "The call of conscience is an appeal to the 'they-self' [Man selbst] in the self; as such an appeal, it summons the self to its potentiality-for-Being-its-self, and thus calls Dasein forth to its possibilities" (SZ 274, MR 319). The call of conscience emanates from the possibility of authentic existence in us.

But this indicates that authentic existence is possible only in so far as we "will to have conscience" (SZ 288). Heidegger thus ties the old problem of the will to that of conscience—disentangling both from "morality" and treating them as ontological determinants. The will can be modified just as the whole structure of Dasein can. Inauthentic will does not want to hear the call coming from futurity; authentic will gives itself over to that call. Inauthentic will sticks to a few possibilities; authentic will is radical openness to any possibility, since it is openness to the ultimate possibility, i.e. death.

This understanding of Being that runs ahead, does so "resolutely". To will to have conscience is to resolutely let oneself be determined by what lies ahead, and ultimately by the possibility of death. Anticipation and conscience are actually closely tied to such resoluteness. Indeed, resoluteness is anticipation of one's totality.

This connection is spelled out in the form of a question.[99] "What if resoluteness...should reach authenticity only when it projects itself not upon any random possibilities which just lie closest, but upon that uttermost possibility which lies ahead of every factical potentiality-for-Being?" (SZ 302, MR 349). Death as the uttermost possibility is then more than a horizon for factical possibilities. Death "enters undisguisedly into every potentiality-for-Being of which Dasein factically takes hold" (SZ 302, MR 349–50).

Resoluteness is a modification of disclosedness. Care is the general title for the complex disclosedness of Dasein of which resoluteness is a modification. It is also a temporal modification since resoluteness anticipates death. Thus, one can understand that Heidegger calls it "the originary truth of Dasein". But he then says: "In resoluteness we have now arrived at that truth of Dasein which is most originary because it is authentic" (SZ 297, MR 343). This is the core ambiguity of the analysis of authenticity. Dasein's most originary—ontological—structure is anticipation; temporality out of futurity. But at the same time, such a structure can be most originary only "because it is authentic", i.e. ontic. One might understand this the other way around: Dasein can be authentically anticipatory of death because ontologically it always anticipates death. But Heidegger does not say it this way. Anticipatory resoluteness is most originary because it is authentic. To understand authentic totalization further, we have to take a brief look again at that which is to be authentic, namely the self. Dasein's inauthenticity is "a possible mode of Being of its self" (SZ 178).

In the context of *Being and Time* the self cannot be thought of as substance or subject, as that which persists under the many changes affecting Dasein, as a soul, person, etc.[100] For Heidegger, the question of the self is not rooted in the problem of change and permanence, but in that of inauthenticity and authenticity. Authentic totalization "constitutes" the self, making the entire care structure one's own (*eigen*).

There are two lines of argument in *Being and Time* about the originary self. The "they-self" (*das Man*) is one of the many *existentialia,* and it is called an "original phenomenon" (SZ 129). Authenticity here appears as an existentiell modification of the they-self. Here it is the they-self that is primordial, originary. But in other passages, Heidegger says otherwise. We are told, for example, that "inauthenticity has possible authenticity as its *ground*" (SZ 259, JS 239). Here it is authenticity that is primordial, originary, the transcendental condition of inauthenticity.

One can even trace two blatantly contradictory formulations about the priority of either authenticity or inauthenticity. "Authentically Being oneself does not consist in an exceptional state of the subject, a state detached from the they-self, but is rather an existentiell modification of the they-self which is an essential *existentiale*" (SZ 130). Yet, Heidegger later writes: "[t]he they-self...is an existentiell modification of the authentic self" (SZ 317, MR 365). The first position is predominant. Inauthenticity is what we find ourselves in for the most part, and "authentically-being-a-self is shown to be an existentiell modification of the they-self" (SZ 267, JS 247); or: "When Dasein brings itself back from the 'they', the they-self is modified in an existentiell manner so that it becomes *authentic* Being-one's-self" (SZ 268, MR 313); or again: "Authentic existence is nothing which hovers over falling everydayness, but is rather existentially only a modified coming to grips with that everydayness" (SZ 179, JS 167).

Joan Stambaugh, who has come across these contradictions in her translation of *Being and Time,* has tied the priority of inauthenticity to the project of phenomenology, and the priority of authenticity to the method of fundamental ontology. Phenomenology, indeed, is simply meant to "show" what happens in everydayness, and thus the "they" comes first. Fundamental ontology, on the other hand, deals with another kind of priority, with what is the "ground" of man's Being. Thus, one could say that inauthenticity is first in the order of discovery, and authenticity is first in the order of foundation. In the latter order, the real self is prior to the they-self. *Das Man* is the surface, not the core, of Dasein. But the surface is what one "*sees*" first. The core is what "*is*" first.

This brings us back to the ideal of wholeness. Only the authentic self is really a whole. Calculative totalization, as occurs with the they-self, is what is most apparent; but existential totalization is what alone unifies the self. From such originary unification or totalization, fundamental ontology can proceed and discover the "meaning of Being" as temporal directionality or ecstases.

It has already been said that calculative totalization prefigures "calculative thinking",[101] and existential totalization prefigures "letting-be, releasement, and eventually *Ereignis*". Only the authentic self allows beings to be what they are. Thus, properly speaking, only the authentic self is entirely "existing", "ecstatic", "*transcendens*", while inauthentic Being is merely a flight from such "standing-out" of oneself.

Stated otherwise, only in anticipatory resoluteness is Dasein truly ecstatic, does it truly transcend itself. Such authentic transcendence, however, does not abolish the "they" and everydayness.

> Even resoluteness remains dependent upon the "they" and its world... In resoluteness, the issue for Dasein is its own potentiality-for-Being, which as something thrown, can project itself only upon definite factical possibilities. Resoluteness does not withdraw itself from "reality", but discovers first what is factically possible; and it does so by seizing upon it in whatever way is possible for it as its ownmost potentiality-for-Being in the "they".
>
> (SZ 299, MR 346)

Authentic existence is truly "outstanding" because it stands explicitly towards death. It incorporates everyday concrete possibilities into the universal horizon of the possibility of death. Such is authentic totalization. It introduces death into every action. Dasein's dispersion among the "they" is its opportunity for existential totalization: it keeps itself open and free for each and every possibility that may occur.

Projection and thrownness

The second approach to the distinction between authentic and inauthentic Being follows the modalities of projection and thrownness. At a first glance, this is not something particularly novel. Inauthentic Dasein projects itself towards things around it, whereas authentic Dasein projects itself towards death. This is not new, but to develop and understand this distinction between the two modes of transcendence will lead us to clarify further the opposition between authenticity and inauthenticity. It will also indicate how inauthenticity permeates what already in *Being and Time* he calls the mathematical project, and in later writings, technology.

Inauthentic projection and the "mathematical project"[102]

For inauthentic existence it is characteristic to remain absorbed in the individual things at its immediate reach. There were two chief modalities constituting such absorption, both dependent on care: "concern" (*Besorgen*) for things (SZ 56–7, MR 83) and "solicitude" (*Fürsorge*) for another Dasein (SZ 121, MR 157). These two determinations are formal and transcendental, that is, they accompany any mode of Being-with and Being-alongside.

To describe inauthentic solicitude, the inauthentic mode of care for others, Heidegger says: "This kind of solicitude takes over for the other that with which he is to concern himself. The other is thus thrown out of his own position...In such solicitude the other can become one who is dominated and dependent" (SZ 122, MR 158). In inauthentic solicitude, one Dasein is disclosed to another through its involvement in everyday matters. In inauthentic solicitude, the other is not wholly disclosed in his Dasein, but only in this involvement with everyday matters—to formulate it baldly, the other appears as an affair to be dealt with as one deals with things "objectively present". In inauthentic solicitude the other is not "totalized", but reduced, so to speak, to something present-at-hand or ready-to-hand. Solicitude, as inauthentic, turns into concern.

The step from inauthentic solicitude towards inauthentic concern is therefore easy. Solicitude, as inauthentic, ceases altogether to be solicitude. In inauthentic solicitude, Dasein "takes over for the other that with which he is to concern himself" (ibid.). Thus, it does not even encounter the other as other; it only encounters business to be taken care of, deadlines to be met, transactions to be completed—but never the other as a "whole". This kind of solicitude is opposed to authentic solicitude, by which the other is encountered "not to take away his 'care', but rather to give it back to him authentically [eigentlich]" (SZ 122, MR 159).[103] Authentic solicitude encounters the other in his existence, that is, as a whole. "This kind of solicitude pertains essentially to authentic care; it meets the existence of the other, not a '*what*' with which he is concerned; it helps the other to become transparent to himself *in* his care and to become *free* for it" (ibid.).

But not only is inauthentic solicitude reduced to concern, it is reduced to inauthentic concern. It has already been shown that inauthentic existence remains absorbed in things at its immediate reach—but in such a way that these are more or less interchangeable. One is absorbed in family matters, but it would not really make a difference if they were office matters: they are treated the same way. One is absorbed in watching a Western, but it would not really make a difference if it were a soap opera. Here, there seems to be an underlying distinction between a life of job-holders and a life of what Heidegger elsewhere describes as a "calling". For inauthentic concern, and thus also for inauthentic solicitude, everything is the same; everything appears as a matter to be expedited. Thus Heidegger can say—though it is an apparent paradox—that inauthentic care is absorbed in what is closest, but at the same time it "[does not tarry] alongside what is closest...In not tarrying, [it] is concerned with the constant possibility of *distraction*" (SZ 172, MR 216). On this page, authentic care is then described as "observing beings and marveling at them—*thaumazein*" (SZ 172, MR 216). *Thaumazein,* was the initial attitude from which the project of *Being and Time* arose insofar as it is that attitude that is required for the retrieval of the question of Being.

Inauthentic care, while it is absorbed in what is closest, does not seek the closest, but on the contrary, what is farthest: "There is no longer anything ready-to-hand which we must concern ourselves with bringing close...It tends away from what is most closely ready-to-hand, and into a far and alien world" (SZ 172, MR 216). The opposition close/alien, or near/far, and the very concept of "de-severing" (*ent-fer-nen*) occurs here in a context of inauthenticity, whereas elsewhere it describes the nature of care, and of the understanding of Being itself. "That being which in every case we ourselves are, is ontologically that which is farthest" (SZ 311, MR 359).

Inauthentic solicitude forced us to look into inauthentic concern. This in turn—because of the paradox of absorption and distraction, or nearness and distance—forces us to look into knowledge as inauthentic. There is thus a double reduction operative in inauthentic projection. Projecting the others in inauthentic totalization is to project them as things; projecting things as "all the same" is to project them as numbers—as occurs in the "mathematical project". "What is decisive in the mathematical project is not primarily the mathematical as such, but that it includes an a priori" (SZ 362, JS 331).

In concern, a type of knowledge is always operative. We have already seen the genesis of theoretical knowledge from circumspection (*Umsicht*) and deliberation (*Überlegung*). In any dealing with instruments and equipment, there is a first understanding of the purposeful pattern, of the totality of involvements (*Bewandtnisganzheit*). Of course, this totality of involvements must remain implicit for knowledge to operate in an undisturbed manner. Proximally and for the most part, when knowledge is not an explicit pursuit, we move within such a pattern without any question.

But from what has been said about absorption and distraction, it is clear that inauthentic concern is not engaged in a world that is so familiar. The routine in handling matters of business, of politics, of love, of scholarship, as if they were all "to be handled" alike, is a break with involvement. The world that is distant and alien, which inauthentic concern looks for (for the sake of distraction), is not a world for circumspection and deliberation since the basic familiarity has collapsed. It is a world for calculation, for inauthentic knowing.

Thus, Heidegger steps back from inauthentic solicitude to inauthentic concern to the mathematical project as inauthentic knowledge. In all three cases what is predominant is an indifferent gaze, a kind of look that does not see anything particular—a staring. For such a gaze (which is not the theoretical gaze)[104] everything looks alike and is apparently already known in advance—there is nothing new under the sun. Everything seems familiar because nothing is really familiar, and everything is known because this gaze never cares to really know. What inauthenticity never experiences is the change-over (*Umschlag*) from familiar involvements or circumspection to theoretical knowledge. It holds itself between these two and thus never stands in either of them. But such a change-over was the genesis of science: "circumspective concern with the ready-to-hand changes over into an exploration of what we come across as present-at-hand within-the-world" (SZ 357, MR 409).

In inauthentic projection, the world is totalized by an unending accumulation of the same, in dispersion and distraction; and this happens in such a fashion that the distinctions between equipment ready-to-hand, objects present-at-hand, and any other Dasein, gets blurred. In later writings, this project that unifies all beings according to sheer manipulation and accumulation will be called "technology", whose essence is enframing (*Gestell*).

The sections of *Being and Time* that treat Descartes[105] trace the rise of inauthentic projection phenomenologically, and not historically as one might think. Here inauthentic projection appears as forgetfulness of the world, *Entweltlichung* (decontextualization) (SZ 112), and as the "vulgar understanding of time" (SZ 24). Inauthentic projection is forgetful of the world "because" it understands times as linear. What is worldless and timeless in this manner are mathematical entities. This is developed at the beginning of Section 21. The mathematical project prescribes a definite understanding of Being: "Mathematical knowledge is... the one matter of apprehending beings which can always give assurance that their Being has been securely grasped. Whatever measures up [to the mathematical project] *is* in the authentic sense" (SZ 95, MR 128).

Such supposedly authentic Being is defined according to the concepts of world and time operative in it. "Descartes'...[i]nterpretation and the foundations on which it is based have led him to *pass over* [*überspringen*] the phenomenon of the world"; and "the Being that is accessible in mathematical knowledge...is such that 'it always is what it is'; accordingly, that which can be shown to have the character of something that 'constantly remains', makes up authentic Being" (SZ 95). Notice the ironic usage of the qualification *eigentlich* here—since Heidegger is precisely describing the mathematical project as disregarding both world and time, and as therefore inauthentic.

The treatment of things as "facts", and even more of another Dasein as a "fact", is possible only because the mathematical project flattens the world out into a world of facts: "In this projection, something constantly present-at-hand (matter) is uncovered beforehand...Only 'in the light' of a Nature which has been projected in this fashion can anything like a 'fact' be found" (SZ 362, MR 414). Heidegger's polemic against science does not go so far as to say that science, technology, and for that matter philosophy since Descartes are all together inauthentic. They do spring, however, from a project that is forgetful of world and time, forgetful of death, and thus unable to totalize "facts" into "phenomena". Science, technology, metaphysics, and logic would thus become integrated into an authentic project if we took a new attitude towards the world and time, *that is,* towards death.[106]

It would be redundant to say that the domain of inauthenticity falls apart into discrete entities—into "detached products" [abgelöste Produkte] (SZ 177, MR 221).[107] Equipment, or logical "assertions", technological products, human affairs etc. are heteromorphous entities not unified by an existential project. But such self-identity, or "exact definability", is the very project of mathematics. Mathematics as a world-project is the condition for totalization by aggregation.

What is definable, individualized, "exactly" known, is also that which can be tabulated, calculatively totalized. The mathematical project isolates facts in the world, and subsumes them under general laws. Subsumption under universals is the kind of "knowledge" that is at the bottom of inauthenticity. By such subsumption, a fact is made objective. When such subsumption is not possible, a fact is not knowable. Facts are knowable by their specific difference. But that was the starting point of *Being and Time*. Definition by genus and specific difference will never allow for an understanding of Being that retrieves it as an issue. Definition by subsumption under laws only applies to beings "present at hand" or objectively present. The extension of that domain of things into all other domains is what is properly inauthentic.

Authentic projection[108]

Inauthentic Dasein understands itself out of *Vorhandenheit*. It remains entirely "thrown" into dispersion. Authentic Dasein, on the other hand, transcends what is given or factual. Therefore, Being, which for us is always an issue, can be called the "*transcendens* pure and simple". Now, in the temporal analysis, such self-transcendence has been called "anticipatory resoluteness" (*vorlaufende Entschlossenheit*). Thus, authentic Dasein understands itself not as something objectively given, but as an event (*Geschehnis*). It *can* understand itself that way because ontologically it *is* always already that way.

Authenticity diminishes dispersion. Authentic projection establishes relations which integrate beings around us into a "world". This world has a dual character due to the distinction between thrownness and projection. Dasein is "thrown" into the world while at the same time "projecting" a world. Authentic projection makes this retrieval of the world explicit. In this context, then, Heidegger speaks of "beings in their totality" (*Seiendes im Ganzen*), as chosen.

In a world so retrieved, beings lose their dispersed particularity. Here, projection is "gathering". In Section 7B, "The Concept of Logos", the logos is defined as "merely letting something be seen [schlichtes Sehenlassen]" (SZ 34, MR 57). In this section, Heidegger reduces "reason", "speech", "judgment", "concept", etc. to a more originary notion of logos. However, "logos" in *Being and Time* is not yet entirely separated from human affairs. In the later Heraclitus interpretation, logos comes to mean a gathering of things into presence. Heidegger then translates *legein* as "the Laying that gathers [lesende Lege]".[109] In *Being and Time,* the "gathering" remains something that man does, by existing authentically. In the later writings, the laying that gathers things into an order of presence is seen historically, as epochal—and no longer in relation to man. This is one of the many instances where the transition from *Being and Time* to later texts is a move away from man. Another such instance is the replacement of the concept of Dasein with that of "Thinking"—a thinking that espouses the epochal orders of presence.

Authentic projection gathers things into a *Weltgefüge*. In Section 6, which announces the necessity of a "Destruction of the History of Ontology", it is, in fact, inauthentic projection that is said to require such a destruction: "Dasein has grown up both into and in a traditional way of interpreting itself...Dasein falls prey to the tradition of that which it has more or less explicitly taken hold. This tradition keeps it from providing its own guidance, whether in inquiring or in choosing" (SZ 20–1, MR 41–2). To "destroy" the tradition out of which Dasein always comes to understand itself is thus to free Dasein's inquiry and choice; the destruction allows Dasein to provide its own guidance. In this connection between destruction of ontology and authentic projection, the *three* paths that Heidegger takes in *Being and Time* to recover the "meaning of Being" appear closely conjoined:

1   fundamental ontology as phenomenological hermeneutics, which was origi-
    nally intended to lead to an understanding of Being *as* Time (NI 28);
2   the transcendental analytic of Dasein, which leads to Dasein's Being as tem-
    porality;
3   the historical destruction of ontologies, which leads to a retrieval of the ques-
    tion of Being through time, i.e. through history or "tradition".

Authentic projection ties these three together—at least, that seems to have been Heidegger's intention. Being appears as Time when Dasein explicitly projects itself towards death and thereby destroys the inherited understanding of itself as something objectively given.

The unity of these three elements or strategies is suggested at the beginning of *Introduction to Metaphysics:* "Taking what was said in *Sein und Zeit* as a starting point, we inquire into the '*disclosure of Being*' [*Erschlossenheit von Seins*]. 'Disclosure of Being' means the unlocking of what forgetfulness of Being closes and hides" (EM 15/19). The "disclosure" of the Analytic is now only the starting point for the explicit retrieval, not merely of Dasein's proper self-understanding but of Being itself. Inasmuch as the "unlocking" (*Aufgeschlossenheit*) is an act of authentic projection, the three elements in question appear more closely tied together in *Introduction to Metaphysics* than in *Being and Time:*

- Being *as* Time means that "from the very first sentence [this lecture] strives to depart from the domain" (EM 15/19) in which Being has been understood as that which is constantly present;
- the Analytic of Dasein is the "disclosure of Being" as Dasein's disclosure (SZ 21–2, 37–8);
- the destruction is the "unlocking of what forgetfulness of Being closes and hides".

The distinction between the modalities of Dasein—i.e. of Being-in-the-world, of care, and of temporality—is crucial for the entire project of in *Being and Time* because it allows one to thematize the retrieval of the question of Being as a philosophical explicitation of what implicitly always already happens when Dasein becomes authentic. Thus, there seems to be a two-step movement.

Authentic projection renders the structure of projection explicit, and the project of fundamental ontology renders authentic projection explicit. Fundamental ontology is the discourse, so to speak, of existence having become authentic.

When Heidegger introduces his famous interpretation of *aletheia* in Section 44b, we have to keep in mind these three strategies. "Beings get snatched out of their hiddenness [das Seiende wird der Verborgenheit entrissen]. The uncovering is always, as it were, a kind of *robbery* [Raub]" (SZ 222, MR 265). This is more than a description of what happens in any mental effort, more than the effort of breaking through appearances, of overcoming the natural attitude, or of reaching out toward the essences of things. Indeed, the text continues: "Is it accidental that when the Greeks express themselves as to the essence of truth, they use a *privative* expression—a-letheia?" (SZ 222, MR 265). This kind of phrase, he adds, indicates Dasein's "originary understanding of its own Being" (SZ 222, MR 265). This can only mean, as before, that Dasein does not conquer its own self by casting aside the they-self, that anticipatory resoluteness does not leave everydayness. It means that Dasein does not become authentic by outgrowing the very possibility of inauthenticity, it does not project itself by dismissing thrownness and facticity, etc. In other words: "Being-in-untruth makes up an essential characteristic of Being-in-the-world" (SZ 222, MR 265). In context, this simultaneity of truth and untruth shows "the state of Dasein's Being which we have designated as '*thrown projection*'" (ibid.). But it is clear that all three dimensions mentioned earlier are "aletheiological":

- Being *as* Time is equivalent to "aletheia" as coming-to-be out of nothing; as coming-to-presence out of absence (*phuein*).
- The Analytic of Dasein is where Dasein must "rob", uncover, its own structure and thus make it its own, *eigen, eigentlich,* authentic.
- Destruction means that the history of ontologies is aletheiological in "untruth" since it equates Being with constant presence; but in "truth" because Being as coming to presence has been operative in all ontologies, *immer schon.* It has been remembered, but at the same time covered up.

Authentic projection as thematizing *aletheia* in such a manner is, Heidegger says in *Introduction to Metaphysics,* a battle. "Solely in the enduring struggle between Being and appearance did [the Greeks] wrest Being from beings" (EM 80/105). In appearance the Being of beings is hidden, *lethe.*

When, in later writings, the transition from the Analytic to the question of Time *as* Being is worked out, anticipatory resoluteness as authentic temporality comes to mean a "leap into nothingness". The historical forgetfulness of the question of Being and the thrownness of Dasein have their equivalent in Being as Nothingness: it is no-thing, not comparable to any being. In a narrower context, it was already said in *Being and Time:* "'That nothing ensues' signifies something *positive* for Dasein" (SZ 279, MR 324). (This phrase occurs, in fact, in the analysis of conscience.)

The pairs "authentic projection and thrownness", "*seinsdenken* and metaphysical ontologies", and "Being and Nothingness", are three instantiations of *a-letheia*.

Knowledge and practice

From what has been said about the genesis of the theoretical attitude as well as about the mathematical project as an a priori, it is clear that there is a priority of practice over knowledge in *Being and Time*. By way of conclusion, it is this priority that I should like to explicate a little further.

The starting point of the analytic of Dasein is our daily business with things that we use. As a form of care, this usage is called "concern" (*Besorgen*). Now, it may happen that we cease to be involved with things and, rather, stop to look at them. This is the origin of theoretical knowledge.

> If knowing is to be possible as a way of determining the nature of things objectively present, by observing them, then there must first be a *deficiency* in our concernful involvement with the world. When concern holds back from any kind of producing, manipulating, and the like, it puts itself into what is now the sole remaining mode of Being-in, the mode of just tarrying alongside.
>
> (SZ 61, MR 88)

In other words, concern encompasses both equipment and objects (*Vorhanden*). When concern loses its relation to equipment, only objects remain. Concern with equipment is circumspective; concern with objects is theoretical. The genesis of the theoretical gaze was described as a change-over from one type of concern to the other. But—and this is important—theoretical knowledge is secondary, derived from another type of knowledge, circumspection, which is knowledge only implicitly. Hence, *Being and Time* would contain an argument for the priority of practice over knowledge.

It was shown how *Being and Time* can be seen as the fulfillment of the modern philosophy of subjectivity—fulfillment as epitome, completion, and end. The commentators who read *Being and Time* in this way insist on the reversal of the transcendental status of practice that occurs in the shift of orientation from transcendental subjectivity to Dasein. This is certainly an appropriate way of describing the respective locus of theory and practice in *Being and Time,* but it may not be the entire story. Regarding this reversal, Ernst Tugendhat, who reads *Being and Time* as such a culmination of the tradition of transcendental idealism, says that in *Being and Time* the sequence between theoretical representation and practice "is reversed when compared to Husserl: for an inquiry that begins with objects, an object must first be represented for a (practical) interest in it to arise; Heidegger, on the contrary, asks how disclosedness is at all possible and thus comes to the primacy of the practical".[110]

In other words, Heidegger fulfills the tradition of transcendental subjectivity by bringing the primacy of representation, of cognition, to an end. For transcendental subjectivity, practice must be preceded by cognition. An object must first be constituted by a noetic act of the subject. This allows a practical interest in this object to arise. In Heidegger, things are the other way around. Dasein must first be involved according to some practical interest (protecting itself from cold with a sweater; using a hammer and nails to fix a bookshelf, etc.) in order for a present-at-hand "object" to be possible at all. Thus, for these commentators, Heidegger brings the tradition of transcendental subjectivity to an end—that is, if the "transcendental" subject is understood as being reflective.

The following quote elucidates how theory derives from practice in *Being and Time:*

> The primary kind of dealing...is not bare perceptual cognition, but rather that kind of concern which manipulates things and puts them to use...Such beings are not thereby objects for knowing the "world" theoretically, they are simply what gets used, what gets produced, and so forth.
> (SZ 67, MR 95)[111]

However, in a deeper way, one would have to say that in *Being and Time* the very distinction between theory and practice is no longer adequate and gets dissolved. This point can easily be shown from the later writings, where it is said that "thinking changes the world".[112] But already in *Being and Time* there is a strong polemic against the distinction between theory and practice. Otto Pöggeler writes: "Heidegger uses the term 'care' because he wants to avoid terms such as behavior or practice and prefers to choose a designation that lies beyond the traditional distinction of theory and practice...He speaks to dissolve this traditional opposition between theory and practice".[113]

What cannot be questioned is that cognition receives a derivative status in *Being and Time.* But because Heidegger's relation to inherited concepts is still quite ambiguous here, his vocabulary does not match the novelty of his thinking. Therefore, it is impossible to argue definitively either for *a priori*ty of practice over theory or for a pure and simple dissolution of their difference. Indeed, on this point as well, Heidegger obviously contradicts himself in *Being and Time.* In the key section on this matter, Section 69, we read:

> In characterizing the change-over from the manipulation and using and so forth which are circumspective in a "practical" way, to "theoretical" exploration, it would be easy to suggest that merely looking at beings is something which emerges when concern *holds back* [*sich enthält*] from any kind of manipulation...So if one posits "practical" concern as the primary kind of Being [in the world], then the ontological possibility of "theory" will be due to the *absence* of *practice*—that is, to a *privation.*
> (SZ 357, MR 409)

This was exactly the way the genesis of knowing was described in Section 13, which is entitled "A Founded Mode in which Being-in is Exemplified. Knowing the World". In this section, knowing is said to become possible when "concern holds back from any kind of producing, manipulating and the like" (SZ 61, MR 88). Heidegger actually speaks of a "deficiency" in concerned involvement in the world. Later, he denies precisely this kind of genesis by deficiency, or by "privation" as he then calls it.[114] With "disappearance of praxis", "absence of praxis", "privation", "discontinuance" of usage of tools, we do *not* reach theory. "[T]his is by no means the way in which the 'theoretical' attitude of science is reached" (SZ 358, MR 409).

At any rate, cognition, or theory as scientific, is derivative. But derivative of what? There are two ways of answering this in *Being and Time*. One is derivative of practice (Section 13), the other is derivative of an amalgamate of a certain practice and a certain theory (Section 69). It is clear that Heidegger struggles here to get out of a traditional either-or, that he does not yet possess the terminology to move forward. This weakness is particularly striking in Section 69. The reason why cessation of usage does not by itself lead to theory is that practice is never without theory, and theory never without practice. "Just as practice has its own specific kind of sight [theory], theoretical research is not without a praxis of its own" (SZ 358). What is this praxis that is operative in theoretical, scientific research? "Reading off the measurements which result from an experiment often requires a complicated 'technical' set-up; observation with a microscope is dependent upon the production of 'preparations'" (SZ 358).

Thus, in the only section of *Being and Time* where Heidegger addresses himself to technology (Section 69), he does so in the context of scientific research and the—comparatively harmless—"techniques" of preparing measuring scales or samples for a microscope.[115] The encompassing project of world-transformation that is *later* called "technology", has its antecedent rather in the mathematical project as an existential a priori in *Being and Time*.

Systematically one can therefore speak of a shift in the disjunction between theory and practice from Section 13 to Section 69 (Table 2.7).

One has to add another criticism: at the time of *Being and Time* Heidegger simply did not yet see the essential link between modern science and mastery over the world. The practical elements of science therefore appear as inoffensive, anodyne—like spreading a blood sample on a piece of glass and sliding it under a microscope.

This underestimation of technology in Heidegger's early writings appears in a totally different context in a lecture course given in 1935. He there speaks of

Table 2.7 The disjunction between theory in practice from Section 13 to Section 69

Section 13:	*Zuhanden*/practice	*Vorhanden*/theory
Section 69:	In what is *zuhanden*: practice = manipulation	In what is *vorhanden*: practice = scientific comportment
	Theory = "a specific sight"	Theory = "research"

125

"global technology" and calls it "Americanism", "the rise of mediocrity and sameness in America and Russia". He says that "Russia and America are metaphysically the same, namely in regard to their world character and their relation to the spirit", and "in America and Russia (the practice of routine) grew into a boundless etcetera of indifference and always-the-sameness—so much so that quantity took on a quality of its own" (EM 35/46). One easily recognizes the "mathematical project" as the a priori of inauthentic projection. In 1935, Heidegger still believed that such inauthentic projection could be fought; that "global technology", the American and Russian mathematical project, could be balanced—and now comes the most embarrassing statement in Heidegger's writings—by National Socialism. "The inner truth and greatness of this movement...[consists] in the encounter between global technology and modern man" (EM 152/199). The reason why I quote this line here is that it most strikingly indicates Heidegger's early understanding of global technology as *escapable;* as just one force among others in the twentieth century; as an inauthentic project that can be remedied by an authentic one. That this authentic project then comes to be identified with a particular political movement is another story, the hardest to swallow.

To summarize, in *Being and Time* and other early writings, the distinction between inauthentic projection and authentic projection is spelled out in terms of sameness and otherness. Sameness appears in the quantifying approach to equipment, scientific objects, and human beings, and otherness appears in letting each of these occupy their own region (*zuhanden, vorhande, Mitdasein*). The first description of modern technology arises from this possible modification of the existential structure—but precisely not in terms of the disjunction between theory and practice. This disjunction is worked out in an entirely different context, namely the cessation of daily usage of tools and the emergence of the scientific gaze. The link between these two lines of investigation—the problematic of technology and the problematic of theory and practice—is apparently overlooked in *Being and Time*.

It seems to me that these considerations about the respective status of theory and practice in *Being and Time* are not all that can be said. A different relation holds between the two. This relation is not only such that, phenomenologically, concern precedes theory, but that, within the entire project of *Being and Time,* a certain type of existence is required in order to "think" fundamental ontology.

Let us recall three important points. First, the most originary phenomenon to which the transcendental analytic of Dasein leads is that of temporality. Second, the primary aspect of temporality is the ecstasis called "futurity". Finally, authentic temporality is, for that reason, called "anticipatory resoluteness". But Heidegger, from the beginning of the book onward, makes a certain state of existence the condition for thinking. This state of existence, in the quote from Plato's *Sophist,* was *thaumazein,* wonderment. In the context of temporality, to exist fully in the now-moment of anticipatory resoluteness is the condition for the understanding of temporality. "Dasein becomes 'essential' in authentic existence, which constitutes itself as anticipatory resoluteness" (SZ 323, MR 370). It is true

REINER SCHÜRMANN

that Heidegger never explicitly says that to understand anticipatory resoluteness as the essence of authentic existence one has first to exist in anticipatory resoluteness oneself. Such a reversal of transcendental priorities is worked out only in later texts, and in another vocabulary. But returning to the beginning—*thaumazein* as the condition for the retrieval of the question of Being—we can conclude that authentic existence as an alternative way of understanding our death is the condition for the understanding of Being as time.

Notes

1 Sartre studied in Berlin and Freiburg from 1933–5.
2 This has been stated most clearly in an essay by Schulz, W. (1953/4) "Über den philosophiegeschichtlichen Ort Martin Heideggers", in *Philosophiche Rundschau*, Tübingen: Mohr, pp. 65–93.
3 Ibid., p. 74.
4 Ibid., p. 76.
5 Ibid., p. 80.
6 {Transcendentalism has taught Heidegger to *de-substantialize* Being—thus his re-naming of the book title in the Nietzsche lectures to "Being *as* Time".}
7 Richardson, W. J. (1963) *Heidegger: Through Phenomenology to Thought,* The Hague: Martinis Nijhoff.
8 {"To head toward a star—this only". "The Thinker as Poet", in *Poetry, Language, Thought* (1971) Trans. Albert Hofstadter, New York: Harper and Row, p. 4.}
9 *Wiederholung* could also be translated as "recapitulation" or "repetition".
10 [The manuscript is unclear here as to the subject of the sentence. There is an indecipherable adjective modifying "people".—Ed.]
11 [Please see the Editor's Introduction for a discussion of the capitalization problems regarding the word "being".—Ed.]
12 {Since Parmenides, Plato, and Aristotle we have become *dull*.}
13 {The ontic is to the ontological as the empirical is to the transcendental.}
14 See WG 31b/105b. See also GP 1/1.
15 See also SZ 308, 385–92.
16 {The fundamental ontology of Sections 1–8 of *Being and Time* retrieve the Greek philosophy of Being, whereas the existential analytic in Sections 9–83 offers Heidegger's version of the transcendental philosophy of subjectivity.}
17 Heidegger, M. (1950) *Holzwege,* Frankfurt am Main: Vittorio Klostermann, p. 1. The English translation of this passage is from David Farrell Krell's Introduction to *Early Greek Thinking* (1975) Trans. David Farrell Krell and Frank A. Capuzzi, New York: Harper & Row, pp. 3–4.
18 *The New York Review of Books,* 21 Oct 1971, p. 51.
19 Heidegger, M. (1983) *Aus der Erfahrung des Denkens* (GA 13), p. 81.
20 See Sallis, J. (1978) "Where Does *Being and Time* Begin?" in *Heidegger's Existential Analytic,* ed. F. Elliston, New York: Mouton, pp. 21–43.
21 {"to ti en einai", is "the what a thing was to be".}
22 {The dogma is that ontology is foundational.}
23 {Being "determines".}
24 {"Asking *about*" is directed toward Being; "*questioning*" is a questioning of Dasein; and "*ascertaining*" is directed toward meaning (*Sinn*).}
25 [The manuscript is unclear here. In the original it reads: "(Heidegger says "conceptual determination: 'We are always already involved...')". The above emendation seemed to best reflect the intended meaning, especially as the phrase "conceptual determination"

occurs nowhere in the quoted sentence—Ed.]

26 See Nietzsche, F. (1993) *Thus Spake Zarathustra,* trans. Thomas Common, Part III "The Spirit of Gravity", New York: Prometheus Books, p. 215.

27 See Heidegger, M. (1977) *Vier Seminare,* Frankfurt am Main: Vittorio Klostermann, p. 73.

28 {There is *only one* other possibility, forgetting, i.e. dwelling in implicit pre-under-standing.}

29 {In Aristotle, a science of Being is sought. In its primary sense (first philosophy) the "object" sought is God. In the secondary sense (second philosophy), the "object" sought is nature.}

30 {See the discussion about the tree as metaphor for the relationship between philosophy and science in the Introduction to "What Is Metaphysics?", in *Pathmarks,* ed. William McNeill, 1998, Cambridge: Cambridge University Press, p. 277.}

31 {Sciences are "grounded" in foundational ontologies, which are in turn "grounded" in fundamental ontologies.}

32 {The "*Sein*" in Dasein corresponds to time as *temporalität* and to Being itself. The link between the "*Da*" and the "*Sein*" corresponds to "our being", and to temporality as *Zeitlichkeit.*}

33 See BH.

34 This is Joan Stambaugh's translation of *Geisteswissenschaften* (SZ 38).

35 See Pöggeler, O. (1963) *Der Denkweg Martin Heideggers,* Pfullingen: Günther Neske, p. 67.

36 Ibid., p. 69.

37 Ibid., p. 71.

38 {i.e Socratic *maieutic:* making explicit what we always knew.}

39 {The order of foundation proceeds "tn phusin", i.e. in itself (according to nature). The order of discovery proceeds "pros hemas", i.e. according to us.}

40 See Fürstenau, P. (1958) *Heidegger. Das Gefüge seines Denkens,* Frankfurt am Main: Vittorio Klostermann, pp. 5–17.

41 Concerning the preceding analysis of Dasein as "neutral" there are two conflicting lines of criticism. Sartre denounces it, saying that Dasein is a neuter, it "appears to us as asexual" (Sartre, J. P. [1968] *Being and Nothingness,* trans. Hazel E. Barnes, New York: Citadel, p. 359), as does Levinas when he says, "Dasein...is never hungry" (Levinas, E. [1969] *Totality and Infinity,* trans. Alphonso Lingis, Pittsburgh: Duquesne University, p. 134). Marcuse, on the other hand, praised *Being and Time* for its con-creteness. He called it a movement of thought "structured according to the concrete mean [of those] who have lived and who live in the concrete world...It is wonderful to see, how, from here on, all rigid philosophical problems and solutions are brought back into dialectical motion" (Marcuse, H. [1969] "Contributions to a Phenomenology of a Historical Materialism", *Telos,* No. 4, Fall, p. 16). More recently, of course, Marcuse denounced the "fake concreteness" of *Being and Time* ([1977] *Graduate Faculty Philosophy Journal;* VI[1], p. 31). Rather than asking whether Heidegger's concrete-ness is genuine or fake, it is important to see that *Being and Time* articulates the parameters according to which concrete experiences can be localized. The question "What does it mean for me to be here?" could be given a formal answer in terms of spatial dimensions. But from a compass one does not expect concrete information con-cerning particular streets and roads. The "formal" approach does not constitute a "fake concreteness", but is rather the exposition of the dimensions involved in anything lived concretely. As in Kant, the categories—or *existentialia*—regulate the empirical. Except that here, formalization is phenomenological not logical (as Kant's is, accord-ing, at least, to *Being and Time*).

42 Stambaugh's term for "*vorhanden*". Macquarrie and Robinson use "present-at-hand".

REINER SCHÜRMANN

43 The word "*ursprünglich*" should be translated as "originative" and not "primordial" as in Macquarrie and Robinson.

44 [This quote is actually taken from the section title to §65, which, in both the Stambaugh and Macquarrie and Robinson translations, reads: "Temporality as the Ontological Meaning of Care"—Ed.]

45 See Fürstenau, P. *Heidegger*, op. cit., p. 16.

46 M—This practical philosophy will not be equivalent to an ethics.

47 Macquarrie and Robinson translate "*Befindlichkeit*" (attunement) as "state-of-mind" and "*gleichursprünglich*" (equioriginarily) as "equiprimordially".

48 See Fürstenau, P. *Heidegger,* op. cit., pp. 17–26.

49 [The manuscript supplies no word following "exposed to". There is only a crossed out sentence fragment that reads "finite constellation against which it 'defends' itself", which clearly does not fit into either what proceeds or follows.—Ed.]

50 This correlation between attunement and beings in their singularity is expressed more particularly in "On the Essence of Ground".

51 {As Heraclitus says (fr. 93), "the Lord whose oracle is in Delphi neither indicates clearly [legei] nor conceals but gives a sign (seimainei)", *Heraclitus: Fragments* (1987) Trans. T. M. Robinson, Toronto: University of Toronto, p. 57. The "seimainei" (giving a sign) is equivalent to a hint (*Wink*) in the same way that speech is "Being that can be understood". "The hint...is the message of the veiling that opens up" (US 141/44).}

52 {Equi-originariness should not be thought of in the sense of a "root" but rather in the sense of a "rhizome".}

53 I translate *Befindlichkeit* as attunement and *Stimmung* and *Gestimmtheit* as mood.

54 I translate *Möglichkeit* alternately by "possibility" and "potentiality".

55 See SZ 143 *passim*.

56 {The basic attitude in Heidegger is an "anti-prescriptive" one insofar as it does not offer an "ideal", but rather advocates a "letting-be". Cf. the late Stoic rejection of voluntarism (e.g. tending toward the ideal of the "sage") in favor of letting one's "individual nature" be.}

57 {Later, the limitation of possibilities is formulated in terms of "epochal" limitation.}

58 I translate *Entweltlichung* as "worldless" but it also has the sense of "decontextualization".

59 {Dread, in this sense, functions as a "heuristic of dysfunction".}

60 The historical model for this characterization of Being-in-the-world is quite clear. It comes from Kant's claim that "the conditions of the possibility of experience in general are likewise conditions of the possibility of the objects of experience" (B 197), which is Kant's "highest principle of all synthetic judgments". The difference is that in *Being and Time* this identity of determination is taken out of the context of judgment, of cognition, altogether.

61 {The Self here does not mean the ego (formal "I") but is rather something "at issue". It has traditionally been a *moral* notion, here it is *ontologized.*}

62 See von Herrmann, F. W. (1964) *Die Selbstinterpretation Martin Heideggers,* Meisenheim am Glan: Anton Hain, p. 156.

63 {Cf. the interpretation of man as the "shepherd" of Being in "The Anaximander Fragment"; *Early Greek Thinking* (1975) Trans. David Farrell Krell and Frank A. Capuzzi, New York: Harper & Row, pp. 13–58.}

64 {The Freudian conception of subjectivity is another way of treating Dasein as an entity *vorhanden.*}

65 {What *everyone* says is that "I am different".}

66 "...reflection is nothing but an attention to what is in us"; Leibniz, G. W. (1981) *New Essays on Human Understanding,* trans. Peter Remnant and Jonathan Bennett, Cambridge: Cambridge University Press, Preface, p. 51.

67 [Schürmann notes that, in the sentence "What is decisive is just that inconspicuous domination by others, which has already been taken over unawares by Dasein as Being-with", the "by" is translated as "from" in Macquarrie and Robinson, which Schürmann disputes—Ed.]

68 See Ionesco, E. (1961) *Rhinocéros,* eds Reuben Y. Ellison and Stowell C. Goding, New York: Holt, Rinehart & Winston.

69 [The next sentence in the manuscript reads: "The starting point in everydayness is: not the way philosophers have interpreted the world, but the way Being has been taken away from us", which seems less than clear.—Ed.]

70 Due to this, some translators have rendered *besorgen* as "make provision for".

71 Solicitude *(Fürsorge)*—the modality of care that obtains in the relation to others.

72 It is interesting to note how, in general, in *Being and Time* deficient modes reveal an existential structure. For example, fleeing reveals disclosedness (SZ 185); falling reveals being as an issue for Dasein (e.g. SZ 193); unquestioned identification with the "they" reveals being-towards-death (SZ 255); cowardice reveals dread (SZ 266); things unmanageable reveal readiness-to-hand (SZ 355). This practice of looking at the deficient mode in order to understand a phenomenon is explicitly discussed on pp. 281–6.

73 See Section 2 on the formal structure of the question of Being.

74 See Fürstenau, P. *Heidegger,* op. cit., p. 27.

75 de Beauvoir, S. (1995) *All Men are Mortal,* trans. Euan Cameron, London: Virago.

76 One could quote many instances of this anticipation of death from the literary realm. *Media vita in morte sumus,* said a poem of the Middle Ages. "The cry of birth is the first cry of death" said Kierkegaard. "As soon as a man comes to life, he is at once old enough to die" (SZ 245), said the *Ackermann aus Böhmen,* a popular epic of the end of the fourteenth and beginning of the fifteenth centuries by Johann von Tepla, in which a Bohemian peasant is ridiculed because he protests against death.

77 Cf. Demske, J. (1970) *Being, Man, and Death: A Key to Heidegger,* Lexington: University Press of Kentucky.

78 Meaning = *Sinn* = directionality.

79 See Fürstenau, P. *Heidegger,* op. cit., pp. 28–33.

80 Again, possibility, in *Being and Time,* is not symmetrically opposed to actuality.

81 The translation of *Geschehen* as "historicizing" (as in Macquarrie and Robinson) does not fully capture this sense.

82 *Zukunft* from *kommen.*

83 {The prefix *ge* indicates a cumulative process, a build-up, as in *Gebirge* (mountain range).}

84 Macquarrie and Robinson translate this sentence as "...is possible only by *making* such an entity *present*" (p. 374). Stambaugh has "...only in a *making* that being *present*" (p. 300).

85 For example Harries, K. (1976) "Heidegger as a Political Thinker". *The Review of Metaphysics,* 29(4), June, pp. 642–69.

86 Aristotle, *Physics,* 219a14–220a.

87 Augustine, *Confessions,* XI, 26.

88 See Janssen, P. (1976) *Edmund Husserl. Einführung in seine Phänomenologie,* Munich: Karl Alber, p. 78 and p. 112.

89 See the "Analogies of Experience" in Kant's *Critique of Pure Reason.*

90 {Here we should take "unitary ground" in the sense of something that is *ultimate* without being a *foundation.*}

91 Concrete does not, of course, mean "ontic".

92 {i.e. the three ecstasis "call upon" Dasein, which *responds* in existing.}

93 See Fürstenau, P. *Heidegger,* op. cit., pp. 34–8.

94 See Fürstenau, P. *Heidegger,* op. cit., pp. 34–8.

95 {Compare this to Aristotle who, at the end of Book I of the *Nichomachean Ethics* asserts that one can evaluate a life only at its end.}

96 For this transition, see Heidegger, M. (1998) "Postscript to 'What Is Metaphysics?'", in *Pathmarks,* ed. William McNeill, Cambridge: Cambridge University Press, pp. 231–2.

97 See Fürstenau, P. *Heidegger,* op. cit., pp. 56–60.

98 {In this instance, the heuristic of negativity consists in the "call of conscience", which states "I should not" do as they all do. This leads to an understanding of "possibility" as "what lies ahead".}

99 Phrasing something as a question is often Heidegger's favorite way of stating what he really wants to say.

100 See Stambaugh, J. (1974) "Time and Dialectic in Hegel and Heidegger". *Reasearch in Phenomenology,* vol. 4, pp. 87–97.

101 See Heidegger, M. (1966) *Discourse on Thinking,* trans. John M. Anderson and E. Hans Freund, New York: Harper Torchbooks, *passim.*

102 See Fürstenau, P. *Heidegger,* op. cit., pp. 39–43.

103 Not only "authentically", but "as his own".

104 See SZ 357–8.

105 Sect. 6, 19–21, 43.

106 See, on logic: SZ 129 and 165.

107 See Fürstenau, P. *Heidegger,* op. cit., p. 52.

108 See Fürstenau, P. *Heidegger,* op. cit., pp. 60–1.

109 See "Logos (Heraclitus, Fragment B 50)" in *Early Greek Thinking,* op. cit., pp. 59–78.

110 Tugendhat, E. (1970) *Der Wahrheitsbegriff bei Husserl und Heidegger,* Berlin: Walter de Gruyter, p. 288. [Schürmann's translation.—Ed.]

111 See Prauss, G. (1999) *Knowing and Doing in Heidegger's* Being and Time, Amherst: Humanity Books. See esp. pp. 7–13.

112 Heidegger, M. (1954) *Vorträge und Aufsätz,* Pfullingen: Günter Neske, p. 229. English, *Early Greek Thinking,* op. cit., p. 78.

113 Pöggeler, O. (ed.) (1970) *Heidegger: Perspektiven zur Deutung seines Werks* Cologne: Kiepenheuer & Witsch, Einleitung, p. 34. [Schürmann's translation—Ed.].

114 {The word deficiency is used because in manipulation there is already *Umsicht.* We know "implicitly" that a hammer is not used to make potato salad, but to drive a nail into a board; that it is useless to try to force your foot into a hat, etc.}

115 See Prauss, G. *Knowing and Doing,* op. cit., p. 20.

3

ORIGINARY INAUTHENTICITY—
ON HEIDEGGER'S *SEIN*
UND ZEIT[1]

Simon Critchley

The past beats within me, like a second heart.
John Banville, *The Sea*

Although its author still invites controversy and polemic, and its theses invite much misunderstanding, there is no doubting the originality and massive influence of Heidegger's *Sein und Zeit,* first published in 1927. Some would argue that it is the most important work of philosophy published in the twentieth century. In this lecture, I will attempt to give a reinterpretation of Heidegger's *Sein und Zeit* through an internal commentary on the text in its own terms rather than through some sort of external, strategic and potentially reductive reading.

I will do this by focusing on two phrases that provide a clue to what is going on in *Sein und Zeit: Dasein ist geworfener Entwurf* and *Dasein existiert faktisch* (Dasein is thrown projection and Dasein exists factically). I begin by trying to show how an interpretation of these phrases can help to clarify Heidegger's philosophical claim about what it means to be human. I then try to explain why it is that, in a couple of important passages in *Sein und Zeit,* Heidegger describes thrown projection as an *enigma.* I trace the use of enigma in *Sein und Zeit,* and try and show how and why the relations between Heidegger's central conceptual pairings—state-of-mind (*Befindlichkeit*) and understanding (*Verstehen*), thrownness and projection, facticity and existentiality—are described by Heidegger as enigmatic.

My thesis is that at the heart of *Sein und Zeit,* that is, at the heart of the central claim of the Dasein-analytic as to the temporal character of thrown-projective being-in-the-world, there lies an enigmatic a priori. That is to say, there is something resiliently opaque at the basis of the constitution of Dasein's being-in-the-world which both resists phenomenological description and which is that in relation to which the phenomenologist describes. In the more critical part of the lecture, I try to show with more precision how this notion of the enigmatic a priori changes the basic experience of understanding *Sein und Zeit.* I explore this in relation to three examples that are absolutely central to the argument of Division II: death, conscience, and temporality. I seek to read Heidegger's analyses of each of these concepts against the grain in order to bring into view much more resilient notions of facticity and thrownness that place in doubt the move to existentiality, projection, and authenticity. This is the perspective that I will describe as originary inauthenticity. As will become clear in the course, this line of interpretation has significant consequences for how we might consider the political consequences of Heidegger's work, in particular—and infamously—the question of his political commitment to National Socialism in 1933.

A clue to understanding the basic experience of *Sein und Zeit*

There are two phrases that provide a clue to what is going on in *Sein und Zeit: Dasein ist geworfener Entwurf* and *Dasein existiert faktisch.* That is, Dasein— Heidegger's word for the person or human being—has a double, or articulated structure: it is at once thrown and the projection or throwing-off of thrownness. Yet it is a throwing *off*— which is how I hear the privative *Ent-* in *Ent-Wurf*—that remains *in* the throw. As Heidegger puts it, "*Dasein im Wurf bleibt*" (SZ 179). Dasein is always sucked into the turbulence of its own projection. Dasein is the name of a recoiling movement that unfolds only to fold back on itself. Its existentiality, its projective being-ahead-of-itself, is determined through and through by facticity, it is always already thrown in a world, and in a world, moreover, ontically determined in terms of fallenness: the tranquillized bustle of *das Man* ("the one" or "the they").

This movement of thrown throwing off or factical existence is the structure of *Sorge,* the care that defines the being of Dasein in *Sein und Zeit.* Heidegger summarizes the structure of care with enigmatic formulae, such as "*Dasein ist befindliches Verstehen*" ("Dasein is state-of-minded, or disposed understanding", SZ 260); or again, "*Jedes Verstehen hat seine Stimmung. Jede Befindlichkeit ist verstehend*" ("Every understanding has its mood. Every state-of-mind or disposition understands", SZ 335).

The principal thesis of the published portion of *Sein und Zeit* is that the meaning of care, where meaning is defined as that upon which (*das Woraufhin,* SZ 324*)* the thrown throwing off of Dasein takes place, is temporality (*Zeitlichkeit*). Simply stated, the meaning of the being of Dasein is time. With the term temporality, Heidegger seeks to capture the passage from authentic to inauthentic time and back again. That is, the masterfulness of what Heidegger calls "ecstatic" temporality, consummated in the notion of the *Augenblick* (moment of vision, or blink of the eye), always falls back into the passive awaiting (*Gewärtigen,* SZ 337) of inauthentic time. Thrown projection or factical existing is ultimately the activity of Dasein's temporalizing, its *Zeitigung,* an articulated, recoiling movement, between sinking away in the dullness of the everyday and momentarily gaining mastery over the everyday by not choosing *das Man* as one's hero.

Once this structure begins to become clear, then it can also be seen that thrown projection or factical existing defines the concept of truth. For Heidegger, truth is also a double or articulated movement of concealment and unconcealment that he finds lodged in the Greek term *aletheia.* In Paragraph 44, the famous discussion of truth in *Sein und Zeit,* with an important emphasis that goes missing in the Macquarrie and Robinson translation, Heidegger writes:

> *Die existenzial-ontologische Bedingung dafür, daß das In-der-Welt-sein durch "Wahrheit" und "Unwahrheit" bestimmt ist, liegt in* der *Seinsverfassung des Daseins, die wir als* geworfenen Entwurf *kennzeichneten.*
>
> (SZ 223)

> The existential-ontological condition for being-in-the-world being determined through "truth" and "untruth" lies in *the* [the italics, and hence the linguistic and conceptual force of the definite article is missing in Macquarrie and Robinson] constitution of the Being of Dasein that we have designated as *thrown projection.*

That is, the condition of possibility for the play of truth and untruth in *aletheia* is the claim for Dasein as thrown projection. In his later work, however, Heidegger always wants to read *Sein und Zeit* from the perspective of what he calls "the history of being" (*Seinsgeschichte*) by claiming that the "lethic" element in truth already implies an insight into *Seinsvergessenheit,* the forgetfulness or oblivion of being. Therefore, although Heidegger will admit in his later work that *Sein und Zeit* expresses itself metaphysically, it already implies an insight into the history of being and thereby into what he calls "the overcoming of metaphysics" (*die Überwindung der Metaphysik*). This is how—in a manner that I always find questionable because of the complete assurance with which Heidegger feels himself able to shape and control the interpretation of his work—Heidegger continually seeks to preserve the unity of what he calls his *Denkweg,* his path of thought. To use Heidegger's own idiom from a manuscript on nihilism from the late 1940s, we might say that the basic experience (*die Grunderfahrung*) of *Sein und Zeit* is this belonging together of facticity and existence, of thrownness and projection, of fallenness and surmounting. It remains a hypothesis to be confirmed or disconfirmed by future research as to whether this is the basic experience of Heidegger's work as a whole.[2]

So, what is the being of being human for Heidegger? Or, insofar as the human being is understood as Dasein whose essence lies in *Existenz,* what is the nature of existence? It is care as a temporally articulated movement of thrown throwing off or factical existing. My concern here consists in working out why Heidegger describes this structure as an enigma and what might be the implications of this claim for an interpretation of *Sein und Zeit.* Once the claim for Dasein as thrown projection is introduced in Paragraph 31 on *Verstehen* (SZ 148), which is also where the word enigma makes its most significant entry into *Sein und Zeit,* then the rest of the book is simply the deepening or nuancing of this structure, like a leitmotif in Wagner, moving through a series of variations. Let's call them "enigma variations", to use an English rather than a German example, Elgar rather than Wagner.

What fascinates me in *Sein und Zeit* is what I would call the spinning or oscillating movement of these variations, where Heidegger tries to capture this enigma in a series of oxymoronic formulations: *"Dasein existiert faktisch"*, *"Dasein ist Geworfener Entwurf"*, *"Dasein ist befindliche Verstehen, Jedes Verstehen hat seine Stimmung, Jede Befindlichket ist verstehend"*, *"'Dasein ist in der Wahrheit' sagt gleichursprünglich... 'Dasein ist in der Unwahrheit'"*, etc. ("'Dasein is in the truth' simultaneously says...'Dasein is in the untruth'", SZ 222). As I shall try to make clear presently, the thought that is spinning out or being spun out in *Sein und Zeit* is that of Dasein as the enigma of a temporal stretch, an almost rhythmical movement or *kinesis* of factical existing that is so obvious, so absolutely and completely obvious, that it is quite obscure. As we noted in Chapter 1 with Wittgenstein, "The aspects of things that are most important for us are hidden because of their simplicity and everydayness (*Alltäglichkeit*)".[3]

The enigmatic a priori

The word *Rätsel,* enigma or riddle, kept catching my eye when reading certain key passages from *Sein und Zeit,* so I decided to try and follow its usage systematically. I have found at least 11 places where the words enigma (*Rätsel*), enigmatic (*Rätselhaftig*) and enigmaticity (*Rätselhaftigkeit*) are used in *Sein und Zeit* (SZ 4, 136,137,148 [× 2], 371, 381, 387, 389, 392, 425), and I will examine these below. The word enigma also appears in Heidegger's later work, particularly in his 1942 lecture course on *"Der Ister"*.[4]

Returning to *Sein und Zeit,* in the opening paragraph Heidegger writes: *"in jedem Verhalten und Sein zu Seiendem als Seiendem a priori ein Rätsel liegt"* (SZ 4). That is, in every comporting oneself to beings, or intentional relation to things, there lies an a priori enigma. This claim already begins to strike a rather dissonant note with the formulation of the phenomenological notion of the a priori in the first draft of *Sein und Zeit* in the 1925 *Prolegomena zur Geschichte des Zeitbegriffs* that I discuss in detail elsewhere, where the a priori is that which shows itself in what Husserl calls "categorial intuition". It would seem that the intentional comportment of the phenomenologist directs itself towards, and itself arises out of, something that eludes phenomenological manifestation. This "something" is what I call the *enigmatic a priori.*

However, the form that this enigmatic a priori takes in *Sein und Zeit* becomes much more striking in Paragraphs 29 and 31, on *Stimmung, Befindlichkeit* and *Verstehen.* Heidegger writes that *Stimmung,* mood, brings Dasein to "the That of its There" (*"das Daß seines Da"*) in a way that stares back at it with an inexorable enigmaticity *("in unerbittlicher Rätselhaftigkeit entgegenstarrt",* SZ 136). Let me clarify this point. Heidegger's initial claim in *Sein und Zeit* is that Dasein is the being for whom being is an issue. In Division I, Chapter 5, the claim is that the being that is an issue for Dasein is the being of its 'there', the disclosure of its *Da* (SZ 133). Thus, Dasein is fundamentally characterized by the capacity for disclosure (*Erschlossenheit*). Or, better, Dasein itself *is* the clearing that discloses "*...es selbst die Lichtung* ist...das Dasein ist seine Erschlossenheit" (SZ 133).

As Tom Sheehan points out, this is what Jean Beaufret had in mind in translating Dasein as *l'ouverture,* which we might render as "the open*ed*ness" to convey the idea that Dasein is always already the space of its disclosure.[5] Indeed, rather than thinking of Dasein as being-there as opposed to here, we might think of being-in-the-world as an openedness that is neither here nor there, but both at once.

Heidegger's claim in Paragraph 29 is that the way in which Dasein is its "there" is caught with the notion of *Befindlichkeit,* namely that Dasein is disclosed as already having found oneself somewhere. The means of disclosure for this *Befindlichkeit* is *Stimmung;* namely, that I always find myself in some sort of mood: I am attentive, distracted, indifferent, anxious, bored or whatever. Therefore, Dasein's primary form of disclosure is affective, and this affective disclosure reveals Dasein as *thrown* or delivered over to its existence, its "there". Therefore, what stares inexorably in the face of Dasein is the enigma of its thrownness, the fact that I am, and that I am disclosed somewhere in a particular mood. This fact is like a riddle that I can see but cannot solve.

Perhaps the most thought-provoking usage of enigma in *Sein und Zeit* occurs just a little further on in the text, at the end of Paragraph 31, where Heidegger summarizes the discussion of *Befindlichkeit* and *Verstehen* by introducing the idea of Dasein as thrown projection in a series of sentences that enact the very enigma that is being described,

> *Befindlichkeit und Verstehen charakterisieren als Existenzialen die ursprüngliche Erschlossenheit des In-der-Welt-seins. In der Weise der Gestimmtheit "sieht" das Dasein Möglichkeiten aus denen her es ist. Im entwerfenden Erschließen solcher Möglichkeiten ist es je schon gestimmt. Der Entwurf der eigensten Seinkönnens ist dem Faktum der Geworfenheit in das Da überantwortet. Wird mit der Explikation der existenzialen Verfassung des Seins des Da im Sinne des geworfenen Entwurfs das Sein des Daseins nicht rätselhafter? In der Tat. Wir müssen erst die volle Rätselhaftigkeit dieses Seins heraustreten lassen, wenn auch nur, um an seiner "Lösung" in echter Weise scheitern zu können und die Frage nach dem Sein des geworfenen-entwerfenden In-der-Welt-seins erneut zu stellen.*

(SZ 148)

Let me closely paraphrase rather than translate this passage, as the precision of Heidegger's conceptual expression is difficult to render literally. The first sentence simply summarizes the conclusions of the opening Paragraphs of Chapter 5, namely that the disclosedness of being-in-the-world is constituted through the existentials of *Befindlichkeit* and *Verstehen.* Let's call them (B) and (V). But the following three sentences enact this conclusion in the form of a series of conceptually palindromic statements:

1   In its being-attuned in a mood (B), Dasein "sees" possibilities (V).
2   In the projective disclosure of such possibilities (V), Dasein is already attuned in a mood (B).
3   Therefore, the projection of Dasein's ownmost potentiality-for-being (V) is delivered over to the *Faktum* of thrownness into a there (B).

Enigmatic indeed! But, Heidegger insists, the full enigmaticity (*Rätselfhaftigkeit*) of this enigma must be allowed to emerge, even if this all comes to naught, founders, is wrecked, or shatters into smithereens, which are various connotations of the phrase "*scheitern zu können*". So, although Heidegger adds that out of such a wreckage might come a new formulation ("*erneut zu stellen*") of the question of thrown-projective being-in-the-world, the disruptive force of the enigma is such as to lead to a breakdown over any phenomenological "solution" ("*Lösung*") to the riddle of Dasein.

Turning now to Division II of *Sein und Zeit,* the word enigma appears on the final page of Chapter 4, "Temporality and Everydayness", four times in Chapter 5, "Temporality and Historicality", and once in Chapter 6 on time-reckoning and the genesis of our ordinary understanding of time (SZ 389, 392, 425). I would like to look in detail at one further appearance of enigma, which occurs just after the temporal *Wiederholung* or recapitulation of the analytic of inauthenticity. Heidegger says that Dasein can for a moment—"*für den Augenblick*"—master the everyday, but never extinguish it ("*den Alltag meistern, obzwar nie auslöschen*"). He continues:

> *Was in der faktischen Ausgelegtheit des Daseins* ontisch *so bekannt ist, daß wir dessen nicht einmal achten, birgt existenzial-ontologisch Rätsel über Rätsel in sich. Der "natürliche" Horizont für den ersten Ansatz der existentialen Analytik des Daseins ist* nur scheinbar selbstverständlich.
>
> (SZ 371)

> What is ontically so familiar in the factical interpretedness of Dasein that we never pay any heed to it, conceals enigma after enigma in itself existential-ontologically. The "natural" horizon for the first starting point of the existential analytic is *only seemingly self-evident.*

That is to say, the existential analytic renders enigmatic the everyday ontic fundament of life, what Husserl calls the natural attitude, what Plato calls the realm of *doxa.* But, and this is crucial, Heidegger does not say that the existential analytic overcomes or permanently brackets out the natural attitude of ontic life, it does not achieve some permanent breakout from the Platonic cave.[6] Rather, as Heidegger points out a few lines prior to the above-cited passage, "*Die Alltäglichkeit bestimmt das Dasein auch dann, wenn es sich nicht das Man als 'Helden' gewählt hat*" (SZ 371). That is, even when I have not chosen *das Man* as my hero, when I choose to become authentically who I am, the everyday is not extinguished, it is rather rendered enigmatic or uncanny. That which is ontically so familiar hides enigma after enigma ontologically. Or, in the words of the opening paragraph of the existential analytic, "The ontically nearest and familiar is the ontologically furthest" (SZ 43). The existential analytic of Dasein seems to return ceaselessly to the enigma from which it begins, an enigma which, in Heidegger's words, shatters the seeming self-evidence of any natural attitude from which phenomenology might begin in order to force the philosopher to formulate anew the question of being-in-the-world. That is, Heidegger transforms the beginning point of phenomenology from the self-evidence of the natural attitude to the enigma of a *Faktum,* the fact *that* one is; philosophy begins with the riddle of the completely obvious.

So, my thesis is that at the heart of *Sein und Zeit,* that is, at the heart of the central claim of the Dasein-analytic as to the temporal character of thrown-projective being-in-the-world, there lies an enigmatic a priori, a fundamental opacity that both seems to resist phenomenological description and is that in relation to which the phenomenologist describes. As such, in Kantian terms, we might say that the enigmatic a priori is not only transcendentally constitutive, it is also regulative. It is not only descriptive, or rather a limit to the activity of phenomenological description, but also normative, functioning like an imperative in the philosophical analysis of being-in-the-world. Philosophy must attempt to be equal to the enigma of our being-in-the-world, while knowing all the time that it cannot. My question will now be: what does this fact entail for our reading of *Sein und Zeit?*

How the enigmatic a priori changes the basic experience of *Sein und Zeit*

Heidegger defines "phenomenon" as *was sich zeigt,* what shows itself, and the phenomena that show themselves in *Sein und Zeit* are not empirical facts, but rather the a priori structures of Dasein's being-in-the-world—the existentials (SZ 31). However, if a phenomenon is what shows itself, then an enigma by definition is what does not show itself. It is like a mirror in which all we see is our reflection scratching its chin in perplexity. An enigma is something we see, but do not see through. We might therefore, at the very least, wonder why the vast and sometimes cumbersome machinery of Heidegger's phenomenological apparatus should bring us face to face with an a priori enigma, with a riddle that we cannot solve. We might be even further perplexed that the riddle here is nothing particularly complex, like the final insoluble clue in a tricky crossword puzzle. On the contrary, the riddle here is that of absolute obviousness, the sheer facticity of what is under our noses, the everyday in all its palpable plainness and banality. Yet, it is this riddling quality of the obvious as the very matter or *Sache* of phenomenology that interests me here.

I began by saying that there are two formulae that provide a clue to understanding what takes place in *Sein und Zeit: Dasein existiert faktisch* and *Dasein ist geworfener Entwurf.* Ultimately, I would like to *modify* the way we hear the formulations "thrown projection" or "factical existing" by placing the emphasis on the *thrown* and the *factical* rather than on projection and existence.[7] That is, on my interpretation, Dasein is fundamentally a *thrown* throwing off, a *factical* existing. It should be noted that what is continually appealed to in Heidegger, in *Sein und Zeit* and even more so in the later work, is a change in our capacity for hearing, that is, whether we *hinhören auf* or listen away to *das Man,* or whether we *hören auf* or hear the appeal that Dasein makes to itself (SZ 271—*inter alia Sein und Zeit* can be understood in musical terms, as an immense treatise on sound, hearing and rhythm). It is my hope that a change in the way we hear these key formulae will produce aspect change in the way we understand the project of fundamental ontology.

I will begin to spell out this aspect change presently, but it should first be asked: why is it necessary? It is necessary, in my view, in order to move our understanding of *Sein und Zeit* away from the heroic political pathos of authenticity, consummated in the discussions of fate and destiny in the infamous Paragraph 74 on "The Basic Constitution of Historicity". As Karl Löwith was the first to learn when he met with Heidegger in Rome and Frascati in 1936, although he has subsequently been followed by other scholars, the concept of *historicity* (*Geschichtlichkeit*) is the link between fundamental ontology and Heidegger's political commitment to National Socialism in 1933.[8] Let me try to briefly restate the argument as, prima facie, the connection between historicity and politics will be far from obvious for many readers.

Dasein's authentic anticipation of its death is called "fate" (*Schicksal*) by Heidegger, and this is designated as the originary historicizing or happening (*Geschehen*) of Dasein (SZ 384). Heidegger's claim in Division II, Chapter 5, is that the condition of possibility for any authentic understanding of history lies in Dasein's historicity, which means the self-understanding of the temporal character of being human, i.e. finitude. So, to repeat: the meaning of the Being of Dasein is temporality, and the meaning of temporality is finitude (SZ 331). Dasein's authentic self-understanding of finitude is "fate", and this originary historicizing is the condition of possibility for any authentic relation to history, by which Heidegger means "world historical historicizing" (SZ 19), or, indeed, for any science of history. It is clear that political events, such as revolutions, the founding of a state, or general social transformations, would qualify as world historical events for Heidegger.

Now, it was established in Division I, Chapter 4, that Dasein is always already *Mitsein.* That is, the a priori condition of being-in-the-world is being together with others in that world. As is well known, the everyday, social actuality of this a priori condition of *Mitsein* is called *das Man* by Heidegger, and this is determined as inauthentic because in such everyday experience Dasein is not truly itself, but is, as it were, lived through by the customs and conventions of the existing social world. Now, returning more closely to the argument of Paragraph 74, if fateful, authentic Dasein is always already *Mitsein,* then such historicizing has to be what Heidegger calls co-historicizing (*Mitgeschehen,* SZ 384). An authentic individual life, Heidegger would seem to be suggesting, cannot be lead in isolation and opposition to the shared life of the community. The question therefore arises: what is the *authentic* mode of being together with others? What is an authentic *Mitdasein* that escapes or masters the inauthenticity of *das Man?* Heidegger writes, fatefully in my view: "*Wenn aber das schicksalhafte Dasein als In-der-Welt-sein wesenhaft im Mitsein mit Anderen existiert, ist sein Geschehen ein Mitgeschehen und bestimmt als **Geschick**".* ("But if fateful Dasein as being-in-the-world essentially exists in being-with with others, its historicizing is a co-historicizing and is determined as *destiny.*") So, destiny is the authentic historicizing that I share with others insofar as my individual fate is always already bound up with the collective destiny of the community to which I belong.

Heidegger goes on: "*Im Miteinandersein in derselben Welt und in der Entschlossenheit für bestimmte Möglichkeiten sind die Schicksale im vornhinein schon geleitet. In der Mitteilung und im Kampf wird die Macht des Geschickes erst frei".* ("The fates are already guided from the front in the being-with-one-another in the same world and in the resoluteness for determinate possibilities. The power of destiny first becomes free in communication and struggle", SZ 384.) So, the fates of authentic, individual Daseins are "guided from the front" by the destiny of the collective, a destiny that first becomes free for itself or self-conscious in the activity of communication and struggle.

Obviously, the word *Kampf* has acquired some rather unfortunate political connotations between the period that saw the publication of *Sein und Zeit* and the present. But that is not the worst of it. Heidegger completes this run of thought with the following words: "*Damit bezeichnen wir das Geschehen der Gemeinschaft, des Volkes".* ("In this way, we designate the historicizing of the community, of the people", SZ 384.) So, the authentic communal mode of *Mitsein* that masters the inauthenticity of *das Man* is *das Volk,* the people. In my view, it is the possible political realization of a resolute and authentic *Volk* in opposition to the inauthentic nihilism of social modernity that Heidegger identified as "the inner truth and greatness" ("*der inneren Wahrheit und Größe*") of National Socialism just a few years later in *Einführung in die Metaphysik* in 1935. Despite the horrors of Nazi Germany, Heidegger—to the understandable consternation of the young Habermas writing on Heidegger in his first published essay—stubbornly refused to revise his judgment on "the inner truth and greatness" when the 1935 lectures were published in 1953.[9]

There is, I believe, a systematic philosophical basis to Heidegger's political commitment, which is due to the specific way in which Heidegger develops the concept of authenticity in Division II of *Sein und Zeit* and which culminates in the concept of *das Volk.* That is, the only way in which Heidegger can conceive of an authentic mode of human being-together or community is in terms of the unity of a specific people, a particular nation, and it is the political expression of this possibility that

Heidegger saw in National Socialism in 1933. In other words, as Hannah Arendt obliquely implied throughout her work, Heidegger is incapable of thinking the *plurality* of human being-together as a positive political possibility. Plurality is always determined negatively as *das Man,* as the averageness and leveling down that constitutes what Heidegger calls, between scarce quotes, "publicness" (*"die Öffentlichkeit"*, SZ 127). In my view, the urgent task of Heidegger interpretation—provided, of course, that one is not a Nazi, and provided one is still in the business of thinking, as I do, that Heidegger is a great philosopher—is to try to defuse the systematic link between Heidegger's philosophy and his politics. As should have become clear, the key concept for establishing the link between philosophy and politics is authenticity, and this is what I want to question by developing the notion of what I call *originary inauthenticity,* a possibility of interpretation that is available, if somewhat latent, in *Sein und Zeit.*[10]

Against the heroics of authenticity: evasion, facticity, thatness

Let me try and explain myself by going back to the key concept of *Befindlichkeit:* state-of-mind, attunement, or what William Richardson nicely translates as "already-having-found-oneself-there-ness". Heidegger's claim is that I always already find myself attuned in a *Stimmung,* a mood or affective disposition. Such a mood discloses me as *geworfen,* as thrown into the "there" (*Da*) of my being-in-the-world. For Heidegger, these three terms—*Befindlichkeit, Stimmung,* and *Geworfenheit*—are interconnected in bringing out the nature of facticity. As is well known, Heidegger's early work is a hermeneutics of facticity, a description of the everyday ways in which the human being exists. In being disposed in a mood, Heidegger writes that Dasein is satiated or weary (*überdrüssig*) with itself, and as such its being becomes manifest as a burden or load (*eine Last*) to be taken up. The burdensome character of one's being, the sheer weight of the that-it-is (*Das es ist*) of existence, is something that I seek to evade.

Heidegger writes: *"Im Ausweichen selbst ist das Da erschlossenes"*, or "In evasion itself is the there disclosed" (SZ 135). This is fascinating, because Heidegger is claiming that the being of Dasein's *Da,* the there of its being-in-the-world, is disclosed in the movement that seeks to evade it. Evasion discloses that which it evades. It is precisely in the human being's turning away (*Abkehr*) from itself that the nature of existence first becomes manifest. I find myself as I flee myself and I flee myself because I find myself. Heidegger seems to rather enjoy the paradox, *"gefunden in einem Finden, das nicht so sehr einem direkten Suchen, sondern einem Fliehen entspricht"* ("found in a finding that corresponds not so much to a direct seeking, but to a fleeing", SZ 135). What is elicited in this turning away of Dasein from itself is the facticity of Dasein's being delivered over to itself (*Faktizität der Überantwortung*), and it is this that Heidegger intends by the term thrownness, *Geworfenheit.*

The concept of *Befindlichkeit* reveals the thrown nature of Dasein in its falling movement of turning away from itself. But two paragraphs later in *Sein und Zeit,*

Heidegger will contrast this movement of evasion with the concept of *Verstehen,* understood as ability-to-be, which is linked to the concepts of *Entwurf* (projection) and *Möglichkeit* (possibility). That is, Dasein is not just thrown into the world, it can throw off that thrownness in a movement of projection where it seizes hold of its possibilities-to-be, what Heidegger calls from the opening words of the existential analytic, *Seinsweisen,* ways to be. This movement of projection is the very experience of *freedom* for Heidegger. Dasein is a thrown project—but where Heidegger will place the emphasis on projection, possibility, and freedom as the essential elements in the movement towards authenticity, I would like to propose another possible trajectory of the existential analytic of *Sein und Zeit,* namely originary inauthenticity.

The thought behind the notion of originary inauthenticity is that human existence is fundamentally shaped in relation to a brute facticity or thrownness that cannot be mastered through any existential projection. Authenticity always slips back into a prior inauthenticity from which it cannot escape but which it would like to evade. As we saw above, it is in this movement of evasion, or the self's turning away from itself, that Dasein's embeddedness in factical existence is disclosed. From the perspective of originary inauthenticity, human existence is something that is first and foremost experienced as a burden, a weight, as something to which I am riveted without being able to know why or know further. Inauthentic existence has the character of an irreducible and intractable *thatness,* what Heidegger called above *"das Daß seines Da"*. I feel myself bound to "the that of my there", the sheer *Faktum* of my facticity, in a way that invites some sort of response.

Now, and this is where my proposed aspect change begins to kick in, the nature of this response will not, as it is in Division II of *Sein und Zeit,* be the authentic and heroic *decision* of existence that comes into the simplicity of its *Schicksal* by "shattering itself against death", as Heidegger rather dramatically puts it (SZ 385). The response will not be the heroic mastery of the everyday in the authentic present of what Heidegger calls the *Augenblick* (the moment of vision), which produces an experience of what he calls ecstasy (*Ekstase*) and rapture (*Entrückung*) (SZ 338). On the contrary, the response to the *Faktum* of my finitude is a more passive and less heroic decision, a decision made in the face of a facticity whose demand can never be mastered and which faces me like a riddle or enigma that I cannot solve. As I try to show elsewhere, such a fact calls for comic acknowledgment rather than tragic affirmation.[11]

Dasein is, as Heidegger writes in his extraordinary pages on guilt, a thrown basis (*ein geworfene Grund*). As this basis, Dasein continually lags behind itself: "Being a basis [*Grund-seiend*], that is to say existing as thrown [*als geworfenes existierend*—another of Heidegger's enigmatic formulae], Dasein constantly lags behind its possibilities" (SZ 284). The experience of guilt reveals the being of being human as a lack, as something wanting. In the light of these remarks, we might say that the self is not the ecstasy of a heroic leap towards authenticity energized by the experience of anxiety and being-towards-death. Such would be the reading of the existential analytic—and I do not doubt that this may well have been Heidegger's intention—that sees its goal in a form of *autarky:* self-sufficiency, self-mastery or what Heidegger calls in Paragraph 64, "self-constancy" (*"Die Ständigkeit des Selbst"*, SZ 323). Rather, in my view, the self's fundamental self-relation is to an unmasterable thrownness, the burden of a facticity that weighs me down without my ever being able to fully pick it up. Expressed temporally, one's self-relation is not the authentic living present of the moment of vision, but rather a delay with respect to oneself that is perhaps best expressed in the experience of fatigue or weariness. I project or throw off a thrownness that catches me in its throw and inverts the movement of possibility. As such, the present continually lags behind itself. I am always too late to meet my fate. I would like to think that Heidegger might have had this in mind at the end of *Sein und Zeit* when he writes of bringing us face to face with "the ontological enigma of the movement of historicizing in general" (SZ 389).

It is my hope that if one follows my proposed aspect change from a heroics of authenticity to an originary inauthenticity then a good deal changes in how one views the project of *Sein und Zeit* and its political consequences. My main point is that both aspects are available to an attentive reading, and this is why the young Habermas was right in suggesting that it is necessary to think both with Heidegger and against Heidegger. However, the completion of such a reading is a considerable task whose fulfillment will have to be postponed to the future. In the remainder of this lecture, I would just like to sketch how we might begin this task by briefly examining three central concepts from Division II: death, conscience, and temporality.

Death—the relational character of finitude

First, I think that the notion of originary inauthenticity places in question what Heidegger sees as the non-relational character of the experience of finitude in the death-analysis in Division II, Chapter 1 of *Sein und Zeit.* You will recall that there are four criteria in Heidegger's full existential-ontological conception of death. It is *unbezüglich, gewiß, unbestimmt* and *unüberholbar:* non-relational, certain, indefinite and not to be outstripped. It is only the first of these criteria that I would take issue with, as the other three are true, if banal: it is certain we are going to die; the instant of our death is indefinite, i.e. we don't know when it is going to happen; and it is pretty damned important. However, if the first of the criteria falls, then the whole picture changes.

Heidegger insists on the non-relational character of death because for him, crucially, *"der Tod ontologisch durch Jemeinigkeit und Existenz konstituiert wird"* ("death is ontologically constituted through mineness and existence", SZ 240). Therefore, dying for another (*sterben für*) would simply be to sacrifice oneself (*sich opfern)* for another, or to substitute (*ersetzten,* SZ 239) myself for another. Thus, the fundamental experience of finitude is non-relational, and all relationality is rendered secondary because of the primacy of *Jemeinigkeit.*

Now, I think this is just wrong. It is wrong empirically and normatively. I would want to oppose it with the thought of the *fundamentally relational character of finitude,* namely that death is first and foremost experienced in a relation to the death or dying of the other and others, in being-with the dying in a caring way, and in grieving after they are dead. Yet, such relationality is not a relation of understanding: the other's dying is not like placing an intuition under a concept. It is not a relation of subsumption, in Kantian terms a reflective rather than a determinate judgment. In other words, the experience of finitude opens up in relation to a brute *Faktum* that escapes my understanding or the reach of my criteria. Deliberately twisting Heidegger's example from Paragraph 47, I would say that the fundamental experience of finitude is rather like being a "student of pathological anatomy" where the dead other "'*ist ein* lebloses *materielles Ding*" ("a *lifeless* material thing", SZ 238). With all the terrible lucidity of grief, one watches the person one loves—parent, partner or child—die and become a lifeless material thing. That is, there is a thing—a corpse—at the heart of the experience of finitude. This is why I mourn. Antigone understood this well, it seems to me, staring at the lifeless material thing of her dead brother and demanding justice. Authentic Dasein does not mourn. One might even say that authenticity is constituted by making the act of mourning secondary to Dasein's *Jemeinigkeit.* Heidegger writes, shockingly in my view, "We do not experience the death of others in a genuine sense; at most we are just 'there alongside' *(nur 'dabei')*" (SZ 239).

If death and finitude are fundamentally relational, that is, if they are constituted in a relation to a lifeless material thing whom I love and this thing casts a long mournful shadow across the self that undoes that self's authenticity, then this would also lead me to question a distinction that is fundamental to Heidegger's death-analysis. Heidegger makes the following threefold distinction:

1   dying, *Sterben,* which is proper to Dasein; which is the very mark of Dasein's ownness and its possibility of authenticity;
2   perishing, *Verenden,* which is confined to plants and animals; and
3   demise, *Ableben,* which Heidegger calls a *Zwischenphänomen* between these two extremes, and which characterizes the inauthentic death of Dasein (SZ 247).

Now, although one cannot be certain whether animals simply perish—"if a lion could talk, we could not understand him"—I have my doubts, particularly when one thinks of domestic pets and higher mammals. Thus, I think one should at the very least leave open the possibility that certain animals die, that they undergo *Sterben* and not just *Verenden.* I also doubt whether human beings are incapable of perishing, of dying like a dog, as Kafka's fiction and the facts of famine, war and global poverty insistently remind us. And what of those persons who die at the end of a mentally debilitating disease, or who die while being in what is termed "a permanently vegetative state"? Do they cease to be human on Heidegger's account? I see no other option. But, more importantly, if finitude is fundamentally relational, that is, if it is by definition a relation to the *Faktum* of another who exceeds my powers of projection, then *the only authentic death is inauthentic.* That is, on my account, an authentic relation to death is not constituted through mineness, but rather through otherness. Death enters the world not through my own *timor mortis,* but rather through my relation to the other's dying, perhaps even through my relation to the other's fear, which I try to assuage as best I can.

It is this notion of an essentially inauthentic relation to death that both Maurice Blanchot and Emmanuel Levinas have in mind when reversing Heidegger's dictum that "death is the possibility of impossibility" into "death is the impossibility of possibility" (SZ 262). I have power neither over the other's death nor my own. Death is not a possibility of Dasein, but rather describes an empirical and normative limit to all possibility and to my fateful powers of projection. My relation to finitude limits my potentiality and my ability to be *(Seinkönnen).* In my view, the experience of finitude impotentializes the self and disables the healthy virility of authentic Dasein.

Conscience—undoing the self

Once this relational picture of finitude is in place, the picture of conscience would also have to change significantly. I have come to think—against some long-held prejudices about Division II—that the discussion of conscience is one of the most explosive and interesting parts of *Sein und Zeit,* and we have already had occasion to discuss certain passages above. Of course, the analysis of conscience follows on logically from the death analysis, being the concrete ontic-existentiell testimony or attestation *(Zeugnis,* SZ 267) for the formal ontologico-existential claim about death. Death is ontological, conscience is ontic. Indeed, the word testimony might detain us more than it has done in reading *Sein und Zeit.* Testimony evokes both a notion of witnessing as testifying to something or someone, and also expresses a link to evidence and verification, where Heidegger is seeking in conscience the concrete ontic evidence for the formal ontological claim about death, a question that resolves itself relativistically in the key concept of "Situation" (SZ 299–300).

My point here is simple: if death is non-relational for Heidegger, then also, *a fortiori,* conscience is non-relational. Heidegger writes, in italics: "*In conscience Dasein calls itself*" ("*Das Dasein ruft im Gewissen sich selbst*", SZ 275). That is, although in conscience it is as though the call of conscience were an alien voice *(eine fremde Stimme,* SZ 277) that comes *über mich,* such a call, although it is not planned, really comes *aus mir.* Its source is the self. As Heidegger insists in differentiating his concept of conscience from the "vulgar" one, what is attested to in conscience is Dasein's ownmost or most proper ability to be *(eigensten Seinkönnen,* SZ 295). Authentic Dasein calls to itself in conscience, and it does this not in the mode of chattering to itself, but rather in discretion *(Verschwiegenheit)* and silence *(Schweigen).* This behavior is what Heidegger calls resoluteness *(Entschlossenheit),* which is then defined as the "*authentic Selfhood*" of Dasein (SZ 298). Heidegger completes this train of thought in a slightly troubling fashion by claiming that when

Dasein has authentically individuated itself in conscience, "...it can become the 'conscience' of others [*zum 'Gewissen' der Anderen werden*]. Only by authentically being-their-selves in resoluteness can people authentically be with one another..." (SZ 298). Once again, the condition of possibility for collective authenticity or community is the mineness of individual conscience.

This brings me to my question: is conscience non-relational? It would seem to me that a consideration of Freud might throw some helpful light on Heidegger's concept of conscience.[12] The Freudian thought I would like to retain is that of conscience as the psychical imprint, interior mark, or agency, for a series of transferential relations to the other: ego ideal, paternal super-ego, maternal imago, or whatever. Conscience is the *Über-Ich* that stands *über mich*, it is the super-ego that stands over against me. The point is that a Freudian concept of conscience is essentially relational. Furthermore, in analytic experience it is the place of the hostile super-ego that the analyst has to occupy in order to break down the symptom that is the occasion of the patient's suffering. Conceived in this way, the appeal made by conscience would not be Dasein calling to itself, or even the voice of the friend that every Dasein carries within it (SZ 163). If that were so, then Dasein would have to be its own best friend, which is a rather solipsistic, indeed slightly sad, state of affairs. Even worse, I would want to avoid Heidegger's suggestion that the authentic self can become the conscience of others in some sort of presumptuous and potentially dominating way.

On my picture, conscience is the ontic testimony of a certain splitting or undoing of the self in relation to a *Faktum* that it cannot assimilate, the lifeless material thing of the experience of mourning and grief that the self carries within itself and which denies it from achieving self-mastery. It is this failure of autarky that makes the self relational. The call of conscience is a voice within me whose source is not myself, but is another's voice that calls me to respond. Pushing this slightly further, the relational experience of conscience calls me to a responsibility for the other that one might consider ethical. In other words, a relational and arguably ethical experience of conscience only becomes possible by being inauthentic, that is, in recognizing that I am not the conscience of others, but rather that it is those others who call me to have a conscience.

It would here be a question of reading Freud's concept of narcissism, as a splitting of the self into conflictual agencies (the division of ego, super-ego, and id in what is usually called the second topography), back into *Sein und Zeit*. If authentic Dasein cannot mourn, because its fundamental relation to finitude is a self-relation, then I think this is because, to put it in psychoanalytic terms, it has not entered into the relational experience of transference. Transference is a relation to another whom I face, but whom I cannot completely know, whom my criteria cannot reach. Such a face-to-face relation is described by Levinas with the adjective "ethical". Of course, *Mitsein* is being-with-another, but it is standing shoulder-to-shoulder with those others in what Heidegger calls in one passage "**eigentliche** *Verbundenheit*" ("*authentic* alliance or being-bound-together", SZ 122). Such alliance might well be said to be the camaraderie that induces the political virtue of solidarity, but it is not a face-to-face relation, and as such, in my view, is ethically impoverished. I sometimes think that authentic *Mitsein* is a little like being in church, it is a congregational "being-together-with-others" where we vibrate together as one body in song and prayer. Pleasant as it doubtless must be, such is not the only way of being with others.

Temporality—the primacy of the past

If we begin to hear thrown projection as *thrown* projection, and factical existence as *factical* existence, then I think Heidegger's claims about temporality—the very meaning of being—would also have to be revised, away from the primacy of the future and towards the primacy of the past. To recall, Heidegger's claim in his discussion of temporality is that there are three "ecstases" of time: the future (*Zukunft*) that is revealed in the anticipation of death, the past or "having-been-ness" (*Gewesenheit)* that is opened in the notion of conscience, guilt and resoluteness, and the present or "waiting-towards" *Gegen-wart*) that is grasped in the moment of vision (*Augenblick*), or taking action in the Situation. The claim is that Dasein *is* the movement of this temporalization, and that this movement *is* finitude: "*die ursprüngliche Zeit ist endlich*" ("primordial time is finite", SZ 331).

Now, although Heidegger insists that the structure of ecstatic temporality possesses a unity, the primary meaning of temporality is the future (SZ 327). As Heidegger writes: "*Zeitlichkeit zeitigt sich ursprünglich aus der Zukunft*" ("temporality temporalizes itself primordially out of the future", SZ 331). That is, it is the anticipatory experience of being-towards-death that makes possible the *Gewesenheit* of the past and the *Augenblick* of the present. For Heidegger, the *Augenblick* is the authentic present that is consummated in a vision of resolute rapture (*Entrückung,* SZ 338), where Dasein is literally carried away (*ent-rückt*) in an experience of ecstasy.

Rapture, which we encountered above in the discussion of authentic *Mitsein,* is a word that worries me, not least because of the way in which *raptus* seems like a plundering of the past, some sort of rape of memory. If we approach *Sein und Zeit* in terms of the aspect change that I am proposing, and we emphasize the thrown-ness in thrown projection and the facticity in factical existence, then this would entail the primacy of the past over the future. This past is one's rather messy, indeed often opaque, personal and cultural history. In my view, it is this personal and cultural thrownness that pulls me back from any rapture of the present into a lag that I can never make up and which I can only assemble into a *fate* on the basis of a delusional and pernicious notion of historicity, and into a *destiny* on the basis of a congregational interpretation of that delusion.

On the contrary, from the perspective developed in this lecture, the unfolding future always folds back into the experience of an irredeemable past, a past that constitutes the present as always having a delay with respect to itself. Now is not the now when I say "now". My relation to the present is one where I am always trying—and failing—to catch up with myself. As such, then, I do not rise up rapturously or affirmatively into time, becoming as Nietzsche exclaimed on the verge of madness, "all the names in history". No, I wait, I await. Time passes. For Heidegger, this is the passive awaiting (*Gewärtigen*) of inauthentic time. Of course, such a passive awaiting might make the self fatigued, sleepy even. As such, in the experience of fatigue, the river of time begins to flow backwards, away from the future and the resolute rapture of the present, and towards a past that I can never make present, but which I dramatize involuntarily in the life of dreams. True, I can always interpret my dreams or, better, get another to interpret them for me. But what Freud calls the navel of the dream, its source, its facticity, always escapes me, like an enigma.

Conclusion

In closing, let me try to identify three consequences that can be drawn from the reading of *Sein und Zeit* which I have tried to offer in this lecture.

1   The critique of authenticity, particularly with regard to social and political life, permits a revalorization of inauthentic social existence as something to be judged positively and not seen in terms of categories such as falling. Although Heidegger insists again and again—indeed, the man doth protest too much, methinks—that his concepts of falling, thrownness and inauthenticity do not and should not imply any moral critique of the modern world, there is no way around the feeling of Heidegger's lip curling as he describes the leveled down life of the "great mass", or—in some twisted echo of Lenin or Kautsky—the "real dictatorship of the 'they'". Such a dictatorship is evidenced in the life of leveled down "publicness", in reading "newspapers" and using "public transport", where "...every Other is like the next" (SZ 126–27). However, if we view Heidegger's descriptions from the perspective of originary inauthenticity, then a good deal changes. For example, when Heidegger writes that in the world of "the they" "...everyone is an other and no one is himself", or indeed when he says that the "who" of everyday Dasein is *Niemand*, nobody, then such phrases might be otherwise interpreted (SZ 128). If we are, indeed, others to ourselves in social existence, if we are even nobody in particular, then this could well provide the basis for a thinking of sociality that would not be organized in terms of the goals of authenticity, autarky, or communitarian solidarity. Reading *Sein und Zeit* from the perspective of inauthenticity might allow us to see social life as constituted in relations of radical dependence on others. I am nobody in particular and nor are you, and insofar as we are both using "public transport" to get to work, then our interactions are based on a shared dependence and even civility. I might pick up and read the "newspaper" that you leave on the seat and we might even exchange a few courteous words about the dreadful terrorist explosion that happened the previous day (I first wrote that sentence on 8 July 2005, on the morning after the terrorist attacks in central London). The point is that *das Man* need not be seen as an inauthentic or leveled down

"publicness" that requires the authenticity of *das Volk*. We might simply abandon the latter and affirm the former. This leads to my second point.

2   On my interpretation of *Sein und Zeit*, the core of the existential analytic is not the heroic, non-relational and constant self who achieves authentic wholeness through anticipatory resoluteness. On the contrary, sociality begins with an inauthentic self who is formed through a relational experience of finitude and conscience. This is not an autarkic and unified self that can rise up to meet its fate, but a self defined through its relations of dependence on others, a self that only *is* through its relations to others and which always arrives too late to meet its fate; it is a comic rather than a tragic self. Indeed, such an essentially inauthentic self might not enjoy the robust health of authentic Dasein; it might be uneasy with itself, even unwell (the possibility of a sick Dasein never seems to have occurred to Heidegger). Such a self might be less an individual than a "dividual", divided against itself in the experience of conscience. In a key passage from the analysis of *Mitsein*, Heidegger writes: "...because Dasein's being is being-with, in its understanding of being already lies the understanding of the Other [*das Verständnis Anderer*]" (SZ 123). For Heidegger, the relation to the other is based on understanding, whose condition of possibility is the understanding of being. However, if we privilege the inauthentic and relational self, then this is not a self that can claim to understand the other, but is rather a self who is directed towards the other in a way that is neither based in understanding nor culminates in understanding. Perhaps the other person is simply that being that I have to acknowledge as refractory to the categories of my understanding, as exceeding my powers of projection or the reach of my criteria, but together with whom I am thrown into a social world where we can engage with each other based on relations of respect and trust. Perhaps it is such an inauthentic self that is truly ethical.

3   The temporality of such a relational self would not be primarily oriented towards the future, a future that culminates in the rapturous "moment of vision" through what I see as a redemption of the past, understood as one's "having-been-ness". On the contrary, such an inauthentic, relational self would be organized in relation to a past for which it is responsible, but which it cannot redeem, a past that constitutes the self without the self constituting or reconstituting it. It is in this way, perhaps, that we might be able to push the existential analytic towards the issue of responsibility for the past, even a guilty responsibility for a past that cannot be fully made present and which, for that very reason, will not let go and cannot be passed over in silence.

What I hope to have done in this lecture is to begin to think about how we might approach Heidegger's existential analytic in a way that frees it from what I see as its tragic–heroic pathos of individual and collective authenticity, but in a way that is hopefully not based on a strategic or reductive external interpretation, but a possible internal reading that derives from the central theses and basic experience of *Sein und Zeit*.

Notes

1 The following text was used as a basis for a lecture course on Heidegger's *Sein und Zeit* at the New School for Social Research in Fall 2005. A greatly shortened version of the text was presented at the American Philosophical Association, Eastern Division, December 2005, "Perspectives in Analytic and Continental Philosophy". I'd like to thank Stephen Watson for the invitation and Robert Brandom and Christina Lafont for their responses. An earlier and rather different version of this text was published in 2002 as "Enigma Variations: An Interpretation of Heidegger's *Sein und Zeit*", *Ratio*, 15(2), June 2002, pp. 154–75. I had long wanted to go back to it as I became unhappy with the formulation of certain of my claims.

2 In this regard, see the interesting *Beilagen* to "Das Wesen des Nihilismus" (1999) *Metaphysik und Nihilismus,* Gesamtausgabe, vol. 67, Frankfurt am Main: Klostermann, pp. 259–67. See esp. pp. 265–6, where Heidegger claims that the essence of nihilism in *Sein und Zeit* is located in the thought of *das Verfallen,* which is the condition of possibility for the surmounting (*Überstieg*) of that fallenness.

3 Wittgenstein, L. (1958) *Philosophical Investigations,* trans. G.E.M. Anscombe, 2nd edn, Oxford: Blackwell, No. 129.

4 Heidegger, M. (1984) *Hölderlins Hymne "Der Ister",* Frankfurt am Main: Klostermann.

5 See Sheehan, T. "Kehre and Ereignis: A Prolegomenon to *Introduction to Metaphysics*", unpublished typescript. See also, Beaufret, J. (1992) *Entretiens avec Frédéric de Towarnicki,* Paris: PUF, p.17, p. 26 and p. 28.

6 This phrase is Rüdiger Safranski's, which he uses to describe the undoubted Platonism of Heidegger's political commitment in 1933. See (1994) *Ein Meister aus Deutschland. Heidegger und seine Zeit,* Munich: Hanser. On the question of the enigma of the everyday in Heidegger see Haar, M. (1989) "L'enigme de la quotidi-eneté", in *Être et Temps de Martin Heidegger: Questions de méthode et voies de recherche,* eds J.-P. Cometti and D. Janicaud, Marseille: Sud, pp. 213–25.

7 *Modifikation* is an absolutely key concept in *Sein und Zeit.* See, for example, the claim that authentic being-one's-self is an *"existentiell modification of the 'they'—of the 'they' as an essential existentiale"* (SZ 130). This is a claim that is simply and flatly inverted in the Second Division, where Heidegger amnesiacally writes, "It has been shown [but where exactly? S.C.] that proximally and for the most part Dasein is *not* itself but is lost in the they-self, which is an existentiell modification of the authentic self" (SZ 317). Is the authentic a modification of the inauthentic, or is it the other way around? Heidegger makes noises of both sorts.

8 See Löwith's essay in (1993) *The Heidegger Controversy* ed. R. Wolin, Cambridge, MA: MIT. To my mind, the systematic connection between fundamental ontology and national socialism was convincingly established by Philippe Lacoue-Labarthe in his "Transcendence Ends in Politics" (1989) in *Typography,* Cambridge, MA: Harvard University Press, and also at greater length in his (1990) *Heidegger, Art and Politics,* trans. C. Turner, Oxford: Blackwell. The same argument has been stated much more polemically and in extraordinary scholarly detail by Johannes Fritsche in (1999) *Historical Destiny and National Socialism in Heidegger's* Being and Time, Berkeley: University of California Press. About the discussion of historicity, Fritsche claims: "...Section 74 of Heidegger's *Being and Time* is as brilliant a summary of revolutionary rightist politics as one could wish for" (p. xii).

9 See Heidegger, M. (1953) *Einführung in die Metaphysik,* Tübingen: Niemeyer, p. 152. See also Habermas, J. (1953) "Mit Heidegger gegen Heidegger denken. Zur Veröffentlichung von Vorlesungen aus dem Jahre 1935", *Frankfurter Allgemeine Zeitung,* 25 July, pp. 67–75.

10 Let me add that I find it curious, to say the least, that certain interpretations or borrowings from Heidegger that would want to distance themselves decisively from any stain of National Socialism often deploy the concept of authenticity in an unquestioned manner. In my view, this is somewhat problematic. I am thinking in particular of the work of Charles Guignon (see *On Being Authentic*, 2004, London and New York: Routledge) and Charles Taylor (see *The Ethics of Authenticity*, 1992, Cambridge, MA: Harvard University Press).

11 See the final chapter of my *On Humour* (2002, London and New York: Routledge), pp. 93–111; and "Displacing the Tragic–Heroic Paradigm in Philosophy and Psychoanalysis" (1999) *Ethics, Politics, Subjectivity,* London and New York: Verso, pp. 217–38.

12 See Freud (1984) "On Narcissism: An Introduction", in *On Metapsychology,* The Penguin Freud Library, Vol. 11, London: Penguin, pp. 90–2.

INDEX

a priori 1, 4, 5, 10, 31; as being 30;
  mathematical project 118, 123;
  phenomenological 29–30;
  subjectivization of 29, 30; as temporal
  52n38; universality and necessity 29;
  see also enigmatic a priori
agents 2
aletheia (truth) 11, 34, 64, 67, 75, 122,
  133–4
analytic philosophy 48
Anglo-American philosophy 45
anticipatory resoluteness 105, 113–15,
  116, 122, 126–7
aporia 62–4
Arendt, Hannah 63, 141
Aristotle 14, 19, 22, 25, 38, 45, 46, 49, 58,
  65, 66, 70, 74, 92, 98
assertion 24–5
attunement 85–6, 87, 91, 109, 110
Augustine, St 98–9, 101
authentic totalization 113–16
authentic–inauthentic distinction 133; and
  anticipation of death 113–15, 116; and
  conscience 114; and originary self
  115–16; and projection 116–23; and
  resoluteness 114–15, 116; and
  thrownness 122–3; *see also*
  authenticity; inauthenticity
authenticity 82, 83, 98–9, 109; against
  heroics of 141–3; and fundamental
  ontology 115; historicity–politics link
  139–41; openness 4–5; and wholeness
  116; *see also* authentic–inauthentic
  distinction; inauthenticity

Bachelard, Gaston 96
The Basic Problems of Phenomenology
  (Heidegger) 12, 33

Beaufret, Jean 136
Being 20–1, 22, 23, 49–50; as aspect of
  phenomenological seeing 10–11; and
  Being itself 62; dogma or prejudice
  concerning 65–6, 67; forgottenness of
  63, 64, 65; foundational character of
  65, 66; general structure of the
  understanding of 83–109; Greek
  articulation of 33; as grounding 69, 71;
  as indefinable 66;
  knowledge/perplexity concerning 66–7;
  meaning of 1, 2–3, 4, 67; originary
  interpretation 82, 84, 88; pre-
  understanding 67–8; preoccupation
  with 58; question of, link with Dasein
  71, 72–3; retrieval of 59–60; as self-
  evident concept 66; and subjectivity
  57; and time 60; topology of 67; truth,
  alethia of 11, 64, 67; twofold priority
  69–73; as universal concept 65–6;
  wood–paths metaphor 63–4
Being In the World (Dreyfus) 1–2
Being as Nothingness 122
Being as Time 121, 122
Being and Time; against heroics of
  authenticity 141–3; aporia in 62–4;
  being-in-the-world 84–9; care 89–99;
  Cartesian view 1–2; clue to
  understanding basic experience of
  133–5; common thesis 56–8;
  concluding remarks 148–9; conscience
  145–7; Dassein as exemplary being for
  the retrieval 64–83; death 143–5;
  enigmatic a priori 135–8; enigmatic a
  priori changes basic experience of
  138–41; fundamental ontology in
  65–73; general structure of
  understanding of Being 83–109;

realism 26
releasement 5, 113
retrieval 109; and authentic resoluteness
61; continuity with Greek Antiquity 58,
59, 64; Dasein as exemplary being for
64–83; Dasein as point of departure 68;
point of departure 68–9; pre-
understanding 66–8; a priori enigma
60; question of Being 59–60, 64, 80,
83; and *Sinn* 61; something in
ourselves 59; three prejudices 59; and
time 60, 61; what is asked 60; what is
interrogated 60; what is to be
ascertained 60
Richardson, William J. 11, 141
Rimbaud, Arthur 37
Rothacker, Erich 56

Sartre, Jean-Paul 56
Scheler, Max 40, 50
Schelling, F. W. J. 57, 58
scientism 44–7, 48–9, 54–5n75, n85
second nature 45
self 92; and being-in 93–5; and being-with
96–9; is encountered in everyday life
94–5; is modifiable 95; is not given 94;
originary 115, 122; as quest 99;
undoing of 145–7
Sheehan, Tom 136
solicitude 116–18
Sophist (Plato) 58, 59, 63
speech 86, 87, 91, 110
Stambaugh, Joan 115
structural analysis 80–1
structure; authentic totalization 113–16;
inauthentic totalization 112–13
subjectivity 2, 20, 57; and Being 57;
categorial intuition 20, 21; facticity of
57–8; intentionality 12, 17, 31;
ontological foundations 57; a priori 29,
30; subject as "process" 57; subject as
radically autonomous 58;
transcendental 123–4
synthesis 25, 28

Taminiaux, Jacques 11, 12
tautologous formulae 34–7
technology 125–7
temporality 84, 99–109, 119, 126, 132,
133, 147–8, 149; ecstatic as condition
of history 107–9; future, past, present
105–7, 112–13; heuristic function of

death 99–102; mental model 106;
originary 108; physicalist model 106;
as "sense" of care 103–7; totalizing
function of being-towards-death 102–3
thatness 141–3
theory 123–6
theory of forms 24
*The Theory of Intuition in Husserl's
Phenomenology* (Levinas) 41–2
they-self 114, 115–16
thinking 11, 120
thrownness 5, 86–8, 104–5, 109, 122–3,
132, 133–4, 136–7, 138, 139, 141–3
Time 3, 4, 82
Time as Being 122
totality 85, 111; authentic 113–16;
inauthentic 112–13
tradition 34, 39, 41
Trakl, Georg 37
transcendental-phenomenological method
80–83
transcendentalism 71, 76, 83, 108, 121,
123–4, 126; as fundamental ontology
77, 78–9; as stepping back 77; as
stepping beyond 77–8; subject matter
of 79; three elements of 77
truth *see aletheia*
Truth and Method (Gadamer) 45
Tugendhat, Ernst 123

understanding 85–6, 87, 91, 107–9,
110–11; ontic modifications 109–27
universals 26

Von Hermann, F.W. 34–5

Wartenburg, Count Yorck von 67
Weber, Max 49
Whitehead, A. N. 64
wholeness 81–3, 85–6, 88, 90–3, 102, 108,
111, 112, 116
the will 114
Wittgenstein, Ludwig 27, 37, 38, 48
wonder 5
the world 11, 32, 119, 120; *see also*
Being-in-the-world

Related titles from Routledge

Very Little…Almost Nothing: Death, Philosophy, Literature
By Simon Critchley

The "death of man", the "end of history" and even philosophy are strong and troubling currents running through contemporary debates. Yet since Nietzsche's heralding of the "death of god", philosophy has been unable to explain the question of finitude.

Very Little…Almost Nothing goes to the heart of this problem through an exploration of Blanchot's theory of literature, Stanley Cavell's interpretations of romanticism and the importance of death in the work of Samuel Beckett. Simon Critchley links these themes to the philosophy of Emmanuel Levinas to present a powerful new picture of how we must approach the importance of death in philosophy.

A compelling reading of the convergence of literature and philosophy, *Very Little…Almost Nothing* opens up new ways of understanding finitude, modernity, and the nature of the imagination.

Simon Critchley is Professor of Philosophy at the New School for Social Research, New York.

"*Very Little … Almost Nothing* **manages, with some aplomb, to pull off the extraordinarily difficult task of saying something new and interesting about Beckett and Blanchot.**"—*New Formations*

"**Altogether beautifully written, with rich and deep insights. It is the most original and enlightening book I know about the so-called nihilism of present times and its genealogy and a key book for the understanding of the contemporary condition of man.**"—*Michel Haar, Université de Paris*

ISBN 10: 0-415-34048-9 (hbk)
ISBN 10: 0-415-34049-7 (pbk)

ISBN 13: 978-0-415-34048-9 (hbk)
ISBN 13: 978-0-415-34049-6 (pbk)

Available at all good bookshops
For ordering and further information please visit:
www.routledge.com

Related titles from Routledge

Reading Merleau-Ponty: On the Phenomenology of Perception
Edited by Thomas Baldwin

Maurice Merleau-Ponty's *Phenomenology of Perception* is widely acknowledged to be one of the most important contributions to philosophy of the twentieth century. In this volume, leading philosophers from Europe and North America examine the nature and extent of Merleau-Ponty's achievement, and consider its importance to contemporary philosophy.

The chapters, most of which were specially commissioned for this volume, cover the central aspects of Merleau-Ponty's influential work. These include

* Merleau-Ponty's debt to Husserl
* Merleau-Ponty's conception of philosophy
* perception, action and the role of the body
* consciousness and self-consciousness
* naturalism and language
* social rules and freedom.

Reading Merleau-Ponty is an indispensable resource for understanding Merleau-Ponty's path-breaking work. It is essential reading for students of phenomenology, philosophy of mind, and psychology, and anyone interested in the relationship between phenomenology and analytic philosophy.

Contributors: A. D. Smith, Sean D. Kelly, Komarine Romdenh-Romluc, Hubert L. Dreyfus, Mark A. Wrathall, Thomas Baldwin, Simon Glendinning, Naomi Eilan, Eran Dorfman, Francoise Dastur.

Tom Baldwin is Professor of Philosophy at the University of York, UK.

ISBN 10: 0-415-39993-9 (hbk)
ISBN 10: 0-415-39994-7 (pbk)

ISBN 13: 978-0-415-39993-7 (hbk)
ISBN 13: 978-0-415-39994-4 (pbk)

Available at all good bookshops
For ordering and further information please visit:
www.routledge.com

Related titles from Routledge

Routledge Philosophy GuideBook to
Heidegger and *Being and Time*, second edition
Stephen Mulhall

Heidegger is one of the most controversial thinkers of the twentieth century. His writings are notoriously difficult; they both require and reward careful reading. *Being and Time*, his first major publication, remains to this day his most influential work.

Heidegger and Being and Time introduces and assesses

- Heidegger's life and the background to *Being and Time*
- the ideas and text of *Being and Time*
- Heidegger's continuing importance to philosophy and his contribution to the intellectual life of our century.

In this second edition, Stephen Mulhall expands his treatment of scepticism, revises his discussion on death, and reassesses the contentious relationship between the two parts of *Being and Time* with a focus on the notion of authenticity.

This guide is vital to all students of Heidegger in philosophy and cultural theory.

ISBN 10: 0-415-35719-5 (hbk)
ISBN 10: 0-415-35720-9 (pbk)

ISBN 13: 978-0-415-35719-7 (hbk)
ISBN 13: 978-0-415-35720-3 (pbk)

Available at all good bookshops
For ordering and further information please visit:
www.routledge.com